IN SEARCH OF ALASKA'S LOST GOLD MINE

Florine Hirsch

[redacted]

A true story of prospecting adventures and bear encounters in the Alaska wilderness.

IN SEARCH OF ALASKA'S LOST GOLD MINE

A true story of prospecting adventures and bear encounters in the Alaska wilderness.

Library of Congress Preassigned Catalog Card Number: 98-094893

ISBN: 0-9668951-0-x

Hirsch, Florine
In Search of Alaska's Lost Gold Mine

First Printing 1999
Printed in United States of America

IN MEMORY

of

PAUL SOLKA, JR.

Author of

"THE LOST GOLD MINE of the UPPER TANANA"

1908 - 1998

ABOUT THE AUTHOR

Born, Florine St.Onge. Raised in Pelham, NH. Married Robert Hirsch, October 1956. Raised five children. Have ten grandchildren and one great grandson.
Were self-sufficient people lived on a small farm. Owners and operators of Bob Hirsch Lumber for many years.
Moved to Alaska in June 1984. Lived in an old log cabin in Sourdough Alaska for three years. Moved to Tenderfoot Hill near Delta Junction in 1987.
Wrote a weekly series of our Alaska adventures for the Pelham Stateline Express, a local publication in Pelham NH, 1984-1986 and The Salem Observer, Salem NH 1986.
A weekly television series of our Alaska lifestyle called, "Alaska Adventures" and "On the Road Again with the Hirschs" on Pelham Cable Television, Channel 52 since 1994.
At the time of this printing 1999, because of public interest,the programs are being rerun.

CONTENTS

FOREWORD.....ix

ACKNOWLEDGMENT.....x

INTRODUCTION.....xi

CHAPTER ONE.....1

CHAPTER TWO.....16

CHAPTER THREE.....41

CHAPTER FOUR.....58

CHAPTER FIVE.....95

DEDICATION.....113

LIST OF FIGURES

Figure 1: Honda "Big Red" heavily loaded....4

Figure 2: Camp #2....12

Figure 3: Florine crossing rockslide....24

Figure 4: Remote trapper's cabin....33

Figure 5: By-pass bridge....42

Figure 6: Hauling building supplies for camp #3....64

Figure 7: The camp....72

Figure 8: Pedro Dome again....92

FOREWORD

Upon reviewing a final draft of the book, "In Search of Alaska's Lost Gold Mine," by Florine Hirsch I would like to make the following comments:

First, this book is a "must read" for anyone wanting to experience the feeling of a real adventure in the Alaskan bush frontier

Florine was able to capture firsthand between the pages of her book, the fears, the gold prospecting fever and sometimes a weariness that can overcome and won't be held back no matter how great the obstacle. A special talent is required to place the reader right in the middle of the action. Florine has accomplished this excitement and I would highly recommend this book to anyone wanting to read a real life adventure in Alaska today.

Leah Madonna
Owner/editor, A.P. Publishing Co,
Fairbanks, Alaska 99701

ACKNOWLEDGMENT

A true story, written from entries in my daily diary, while searching with my husband, for Alaska's Lost Gold Mine of the Upper Tanana in the Alaska wilderness.

To Tom Vaneyeck and Irene Mead, for their knowledge of the area and interest in the beginning of our search for the lost gold mine that made it possible for Bob and I to continue our adventures.

I owe, Linda Peters, a special debt of gratitude for her interest in the story and encouragement to continue writing.

A warm and sincere thanks to my daughter Debra Leclerc for help and support in preparing this book for publication.

To Dr. James Madonna, Professor Emeritus of Mining Extension at the University of Alaska Fairbanks, I also owe a special debt of gratitude for his teachings, helpful suggestions and time for the help in the completion of this book for publication.

To Leah Madonna a special thank you for all your advise and capturing my true feelings in her written Foreword.

To James Frey, Jr. poet of the poem "Alaska Challenge" for permission to use a portion of his poem.

And a sincere thank you to anyone who has in some way helped me along in the development of this book.

To all our followers through our newspaper articles and television programs in the local area of our hometown Pelham, N.H. and to the many friends and relatives across country who showed much interest in reading of our adventures.

Thank you, one and all!

INTRODUCTION

OUR ADVENTURE BEGAN the summer of 1994, when my husband Bob and I read a book by Paul Solka, Jr. called THE LOST GOLD MINE OF THE UPPER TANANA.

As a young child, the author lived with his parents who were drift miners along Tenderfoot Creek, which is located in our area of Alaska.

A drift mine is a shaft sunk vertically through perma-frost or frozen ground to bedrock, then horizontal drifts branch out once bedrock has been reached, attempting to crosscut the paystreak* of the ancient river bed.

The method used was to take off the moss on the top of the ground and to build a fire with dry wood underneath and cover this with dry logs. Then a fire was lit at the end with the most draft. The cracks between the logs were covered with dirt.

Next morning the unburned logs and water were removed along with the thawed dirt. This was repeated until bedrock had been reached with the miner removing approximately ten inches of thawed dirt per day plus another six inches by picking. The shaft was about two and a half feet wide and usually six feet long. Once a depth was reached when dirt could no loner be thrown on top, a windlass* was used.

This was a two-man operation with one on the windlass hoisting up the dirt andgravel in a bucket. They would begin panning as soon as they reached gravel and continued to do so until bedrock was reached.

In later years a steam point was used to thaw the ground. Gravel from the shaft was transported to a dump on a cabl tram. Then after the spring thaw, this ore carrying gravel was washed through a sluice to remove the gold.

The Offer

One day, while Paul's father and uncle were busy at their own drift mine, a man named Tony, approached them, and asked if they would help him mine in an area rich in gold.

They listened as he excitedly told of an area he found. He did not reveal where it was located except to say it was a long and difficult trip. They agreed to help him when they completed the work on their own drift.

Tony became very possessive of "his territory," the flats of Shaw Creek. Many hours were spent on his horse riding over the surrounding treeless ridges. He tried to keep hunters and trappers away as hunting and trapping were good in this region. The demand for furs by the affluent rich, after World War I, lured many a trapper. From his vantage point on the ridges, he closely watched the mining operations along Eagle and Corda Creeks.

He was known as "Crazy Tony" or "One Eye Tony" because of his disfigured face. A horse had kicked him in the head leaving an ugly scar and a gaping empty socket from the loss of an eye. medical treatment was limited to a nurse who lived in the area. People thought him to be crazy because of his looks and he was soon shipped to Morningside, a state sanitarium in Washington, never to return home.

Several years later at the local store, someone asked the store owner why he had not cleaned the place, like some of the things laying on the windowsill. The owner remembered it had been Crazy Tony who had given him the package wrapped in newspaper to give to someone but he forgot who. Curious, they opened it to find a quartz rock laced with gold! So now the story begins of the search of THE LOST GOLD MINE of the UPPER TANANA.

*Paystreak- concentration of minerals in gravel
*Windlass-apparatus for hoisting

CHAPTER ONE

Partners

Plans for our 1995 excursion started in the fall of 1994 with another couple, Tim and Eva with whom we had become acquainted. Eva, a lifelong Alaskan, grew up in the region and knew of an old trapping trail that would take us to the area that we wanted to go. She estimated it would be approximately twenty-five miles over rough terrain. Knowing this, we decided the four of us would travel together.

Visiting the Lower 48

Winter begins in early October in Alaska and is our signal to make our annual trek back to New Hampshire to visit our family and friends. After two months we bagan our journey back to Alaska, via the southwestern route through Arizona, California, Oregon and Washington.

In early March of 1995, we met with the author, Mr. Solka, at his home in Eugene, Oregon. He told us many interesting stories of his life on Tenderfoot Creek as a young boy and of the mining in the area from 1917 through 1922. When the quartz rock laced with gold had been found at Emard's store, located at the foot of Tenderfoot Hill, he had been there. He verified that the map in his book was a copy of the original map drawn on the back of his mother's calendar by the son of an Indian chief.

Presumably, in 1900, Chief Jarvis had helped two prospectors bring supplies to the area of the lost gold mine. Returning after a few months, he helped one of the prospectors reach the Tanana River with eighty pounds of gold and had been told the other had died a terrible death of a ruptured appendix. So the story is told. Our interest became more intense and we could not wait to return to Alaska to start formulating plans.

Filing Claims

As soon as Tim and Eva learned we had returned to Alaska, they immediately came to visit us, as they too were anxious to start making plans. Remembering that the author had told us he knew someone who was planning to go by helicopter in May to the area, Bob said, "Our first step is to hold the land by filing prospecting sites."

"I did manage to put several of the claim markers up while trapping in the area this past winter," Tim answered.

A prospecting site consists of 160 acres and is good for one year. Except for a recording fee, no rental fee is paid on a prospecting site. A mining claim has 40 acres per claim. The claims are filed and recorded with the State Division of Mining and an annual rental fee is paid to the state or Federal government, depending on location of claims. We each filed and recorded eight prospect sites, which covered two square miles per person, therefore holding a total of eight square miles. We verbally agreed that if gold was discovered on any prospect site or if the lost gold mine was found, all four names would be included when we filed and recorded a mining claim.

Close Call

The week before our first supply trip, as Bob and I were returning home from an evening excursion of trail riding on the three wheelers, I came around the corner of the circular driveway, found myself at the edge of the four foot embankment and over I went. Next thing I knew, Bob was standing over me asking if I was all right. Dazed and shaking I answered, "I think so." He checked me over and except for a sore spot on my left side I seemed fine.

Let me explain about hospitals and clinics here in Alaska. Most small communities such as Delta Junction, Tok, Gakona, etc. including remote native villages have only a small clinic with one doctor, open four days per week. The clinic is closed on weekends and when the doctor is on vacation. For some of us living in the brush outside of a town or village but within the road system, we must be sure when making a decision as to whether we need medical help immediately or not. We do have 911 but it would be some time before help arrived. Now on the other hand, the hospital is located in Fairbanks, seventy miles to the north of where we live and the clinic to the south, thirty miles in Delta Junction.

After a restless night, and with the clinic closed because it was the weekend, Bob checked me over and after much poking for sore spots asked me to take several deep breaths checking for a punctured lung. I had one sore spot on my upper rib. No sharp pain and I could breathe all right, so it must be okay!

Finalizing Plans

We found a difference of opinion after many hours of planning the equipment; camping gear, meals, etc. Eva preferred to take MRE's which are Meals Ready to Eat, sealed in plastic which contain a main meal packet which can be heated in boiling water. Also a powdered drink, dessert, salt, pepper and other items. I, on the other hand, prefer the challenge of cooking over an open fire.

Camping was not a new experience for us as we have traveled most of the trails in the White Mountains of New Hampshire and some in the southwestern United States.

Our list completed, and each person knowing his or her responsibility, we now had to wait for spring to make our first supply trip.

First Supply Trip

We met Tim and Eva at Fish Lake on the morning of May 20th just as it began to sprinkle. Anxious and ready to go we decided to chance it and continued on. In her twenty-foot riverboat, Eva took all the gear, including three fifty-five gallon drums packed with food, plus Tim's dog, Sandy across the lake. The trail the three of us followed, on ATV's, around the lake was not very good which should have been an indication of what was ahead.

Actually, it proved more difficult because of my sore ribs. Yes, I insisted on going. Bob helped me often through some of the more difficult areas and was understanding, knowing I didn't want to miss the first supply trip.

We arrived at the backside of the lake forty-five minutes later to find Eva had already unloaded most of the supplies. To pack the supplies into two trailers and onto three ATV's, proved to be challenging.

Finally, at 9:30 am we started up the trail, traveling very slowly until we knew for sure that everything would stay in the trailers. From the lake the trail went up over a wooded ridge then through a flat boggy area. This bog was several miles long with lots of mud holes, which had to be detoured or driven through. Surprisingly, we had no major problems through this, although the bear droppings I spotted along the trail made me wonder what was ahead of us.

One really needed to be cautious, as these mud holes can be quite deep because of the melting perma-frost. Once the grass or moss has been disturbed, the black dirt absorbs the heat of the sun and begins to gradually thaw the ground creating a mud hole that can be as deep as four or more feet. You should first test the depth before attempting to drive through. As we skirted Rion Lake, we passed an old trapper's cabin, pretty much fallen in. Then at the far end of the lake we watched a magnificent bull moose standing in the water to feed. Close to the shore, near us, was a large beaver house.

After the lake, the trail veered to the left where it crossed a small creek by way of a rickety plywood bridge. Apparently, the trapper who still trapped that area of the trail had built it.

We started to climb as we passed his cabin built next to the trail. Passing through a tall stand of timber we had our lunch by an outcropping of rocks. Within minutes, Sandy had a face full of porcupine quills and some in one paw from trying to catch one. Tim did manage to pull some out, under protest by the dog, but several remained embedded.

Figure 1: Honda "Big Red" heavily loaded for our first supply trip.

Problems

By late afternoon the axle on Tim's trailer had been bent several times caused by carelessness in driving through narrow areas along the trail. He had built the trailer slightly wider than the four wheeler, therefore the tires would hit a clump of dirt, a stump or a tree causing it to bend.

The first few times the pick handle was used to straighten out the axle but eventually that was lost along the trail. Next a black spruce pole was cut because of its strength and was now carried in the trailer. Mind you, this happened several times so we expected it to break at any time.

Trail Worsens

With a heavily loaded trailer, carrying two fifty-five gallon drums, that would rock back and forth while traversing the rough trail, plus other supplies, the Honda Big Red was loaded to its limit and then some. On a few of the smaller steep hills, Tim, Eva and I would push from behind to help Bob get to the top.

Then at the long steep hill of Squaw Peak he hauled one barrel at a time. At the last steep section near the top he ran next to the machine to keep off the extra weight.

After the second trip he returned to drive my machine up while Eva and I decided we would rather walk. We thought of the pioneer women who walked beside the covered wagons on their way west. We felt much safer!

We caught a glimpse of the Tanana River through the brush from the top of the mountain. Also the bluffs, many miles away, along with the hill where our home is located were visible and the two cabins built on Clear Creek could be seen.

Eva found a survey marker dated 1948, but no elevation marked. She said the military had a road to Pedro Dome, a quonset hut and a fire tower on the peak when they were used to spot forest fires. But all was gone now and she thought there was a forestry radio relay tower presently on top of the mountain.

Camp #1

Finally, after seven hours on the rough trail we came upon an old trapper's camp, which became for us, Camp #1. Apparently, the trapper had planned to build a cabin but his first concern was his dogs, which he used for transportation. There were two double, eight-foot long doghouses he had built of spruce logs. The roof was sod and each house had a division in the center to divide the house into two separate sections with an entrance at each end. It was surprising how sturdy they were after sixty years, but they were slowly settling into the ground.

We quickly set up our tents as it began to rain. Bob and I cleared the brush from the flat area where the trapper most likely had cleared for his tent or cabin. The others put theirs under the tarp we had over the cooking area. Later we tried to enjoy a warm campfire after our hearty supper of Dinty Moore stew but it was really getting cold!

During the night, I reluctantly crawled out of the tent to discover it was snowing. Oh no, a heavy wet snow too! Oh well, we were warm and comfortable in our sleeping bags.

The Second Day

A beautiful winter scene surrounded us as we emerged from the tent and into the cold morning air. Everything in sight was blanketed with an inch of heavy wet snow.

When the sun broke through the clouds, the snow on the ground soon melted but the brush remained wet. We wore our rain gear and left the three wheelers at camp to continue to clear the trail. Actually, we were making a new trail as the one we were following had disappeared. By noon, we managed to clear two miles of trail before joining the old trail again.

After a time, we realized we were doing most of the clearing. Eva walked ahead to check the route of the trail while Tim drove the four wheeler through the brush along the trail until he came to an obstruction, like a downed tree. He then waited for us and helped remove whatever was blocking the way. He worked with us for awhile before returning to get the machine then continued past us until the next blocked area. This happened often but we continued on. Bob cut with the chainsaw and I cleared the cut brush from the trail. Even with my sore ribs!

After several hours of clearing trail we decided to check ahead to see what we could expect on our next trip. Tim went ahead of us and we followed more slowly to check the areas on the steep ridges we would have to by-pass. Tim was resting when we caught up to him, then he continued on.

We also decided to rest when all of a sudden Bob had a charley horse in his upper thigh, which would not go away. Most likely it was caused from the chainsaw work and hiking since yesterday. With approximately four miles back to camp, he thought he'd head back down the trail while Eva and I waited for Tim, hoping Bob would at least make the two miles back to the four wheeler.

"Tough to get old," he muttered as he hobbled back down the trail.

Tim returned after twenty minutes. He explained there were several more ridges we must travel over and the mountain was still a long way off.

By then it was late afternoon so we set off for camp. Glancing back from one of the steep ridges we were climbing, we saw the mountain looming high in the distance at 3360 feet. According to our estimate it would be at least eight miles before we reached our destination.

We expected to catch up to Bob, but apparently he made it past the four wheeler. Tim offered me a ride back to camp. Obviously, he was not thinking of the danger of three people riding on the machine while traveling over the rough trail. No thanks, I'd rather walk. Eva insisted I take her 45 Semi-automatic pistol.

"Me? I never shot one yet!" I exclaimed.

"It is a semiautomatic, so whatever you are shooting at, just aim and keep pulling the trigger," she explained.

I nervously strapped the gun and holster around my waist and started back down the trail by myself. One is very conscious and observant of her surroundings and the slightest sound will often startle you. I managed to keep ahead of them even though I was walking. Maybe faster than I thought!

About a mile down the trail I met Bob who was coming toward me on Big Red.

"The charley horse disappeared within the first mile," he said. "Besides, I was worried about my little girl out here all by herself."

"How sweet," I answered with a big smile.

"Get on. I know you're tired from hiking lots of miles," he said.

I did get on but not for long as it was too difficult a trail to ride double. It was easier to walk following behind as he maneuvered the machine over the rough terrain of the trail.

At 3:30 we left Camp #1 after packing supplies in the barrels with locking covers. We then chained them to a tree and hung our tents, pads, and foam, (yes, we did bring a foam pad for extra comfort) in a tall spruce tree, hoping no animal would get to them.

Within ten minutes on the trail, the wheel finally broke at the hub on Tim's trailer. Apparently, it had been bent just too many times on the way in. Bob noted that it had been spot-welded. Tim admitted he hadn't checked it when building the trailer. Now the gear in their trailer was loaded onto our machines and trailer. Even the tire!

Actually, Sandy the dog had been riding in the trailer because of the episode with the porcupine yesterday. He had one swollen, sore paw and now had to walk the fifteen miles back to the lake. My arms and legs were lame from the trip in, plus clearing trail so I drove my machine over the rough terrain while standing to have the strength to steer and control it. I asked Tim to take my machine down the steep section of Squaw Peak. I got on the four wheeler with Eva who kept right on going once we were at the bottom of the mountain. "You need a break. Let Tim drive the three wheeler," she said.

The four wheeler is a nice sturdy machine, but for the cost of $6000, I would manage my $800. three wheeler. We arrived at the lake after a little over three hours. We asked permission of the owner of the Red Spruce Lodge to use their dock to load the boat. Sure, no problem. Bob and Tim loaded the gear and my machine into the boat before taking the trail around the lake. Eva, myself with rather sore ribs, and Sandy who was limping badly after his long walk, crossed the lake. We unloaded what gear we could while waiting for the guys. It started to rain and got real cold, but thankfully we had made it safely back to the vehicles. Good trip!

Next morning, back home it was snowing and stayed at 30 degrees all day which gave us a good excuse to stay warm by the wood stove. It was just as well as we were both lame and, Oh boy, did my ribs hurt. Next trip was planned for June 2nd.

Second Supply Run

After having to cancel several times because of the weather, we finally left on June 17th. It had rained three inches the day before but that morning it was clearing. Eva called at 5:00 am, a prearranged time.

"Let's go. Meet at the lake at 7:00 am." she said.

"Okay!" I answered enthusiastically.

This time, Bob brought 2x6 planks he had notched on one end to hook over the rim of the boat to drive the three wheeler onto it. He did not want to travel over the difficult trail around the lake again. Tim drove his machine over the trail and met us on the backside of the lake.

The first trip across the lake, Bob and Eva took my machine, one heavily loaded barrel, extra gear and the trailer. Next trip included the three of us, Sandy, more gear and Big Red. The boat was full! I was surprised how easily Bob drove Big Red down the planks and into the shallow water, and now understood why he had to load them on backwards, .I was amazed at the amount of supplies we brought in on each trip.

Tim had three weeks in which to fix the wheel for the trailer that had been left out near Camp #1 on our last trip, but hadn't, plus he forgot to bring his chainsaw. With Big Red, Bob pulled the one trailer, again heavily loaded with another barrel and more supplies; five, one gallon cans of gas, a shower can, which was a galvanized sprayer with a spigot Bob had welded near the bottom, plywood and boards for a table top and shelving and one piece with a hole cut for our toilet seat. "For the Dunny!"

On The Trail

The trail was much drier than on the first trip even though we had had lots of rain yesterday. Oh, oh, more fresh bear droppings in the same area as before. I had to convince myself that we hadn't seen a bear on the last trip so why worry! Me? Worried! You bet!

We were making good time until we came to a sign just after crossing the rickety plywood bridge. A sign right in the middle of the trail "NO TRESPASSING PRIVATE PROPERTY"

"No problem, we can by-pass his property and catch the trail further up," stated Eva.

Although this was a public trail, it did not appear to be well used, therefore the trapper whose cabin we had passed on the first trip must have felt his privacy was threatened.

To our left was an old overgrown military road that probably had been bulldozed back in the early 1950s. Traveling through the tall grass was rather difficult but within several hundred feet from the trail we veered off. We had to work our way through the woods and rough terrain to join onto the trail above the cabin. Gee, the trail hadn't improved any since our last trip.

When we approached the steep section of Squaw Peak, Bob asked Tim to pull the trailer, with his machine, to the top only, as Bob preferred to pull his own.

From the top, I tried to call the Red Spruce Lodge on the portable CB we had with us as the owner of the lodge had promised to listen for our call. No luck! We checked from several more ridges but never made contact with anyone. There had to be a way and if there was, I knew Bob, with his Yankee ingenuity would figure something out.

We reached Camp #1 about 1:00 p.m. to discover it had been visited by bears. The bears had left their calling card, known as scat. The tents, sleeping bags etc. hung in the spruce tree had been overlooked most likely because they had no food odor. The barrels were fine but a bear had bitten a plastic gas can as it sure left some large teeth marks. Apparently it didn't like the taste of gas and left it alone. Bob's thirty-year-old backpack had one pocket chewed off. "There was old grease that had leaked in the pack many years ago and evidently that's why it had been chewed," he said

A fire was built and Tim did what he does best, sit and watch it while we gathered the firewood and worked around camp. Eva made an MRE lunch for us, then we packed supplies, including camping gear, onto the machines before continuing on the two mile trail we cleared last time.

Again, Tim had no problem running over small trees with the four wheeler so he and Eva went ahead to check the trail. We continued to clear trail to save abuse of our machines, plus make traveling easier on the way back. After every two thousand feet or so of trail clearing, we returned for the machines, and worked this way all afternoon. At times we spotted fluorescent flagging tape tied to a tree marking the trail as to which way they were traveling. Up one ridge, down another, through a burn area thick with fallen trees and black sooty dirt.

We came upon Tim and Eva about 6:30 p.m. sitting by the four wheeler. Tim asked, "Why did it take you so long? We could hear you coming up and over ridges most of the afternoon."

"Clearing trail," I answered.

"Wasting too much time on trail making," stated Eva. "Besides, we can't find the trail beyond this point and the burn area ahead is thick with fallen tree trunks. Let's return to Camp #1 tonight."

Camp #2

Bob had to see for himself and left to scout the area ahead. Returning, he said, "With a little effort we could work our way through the burn area tonight."

I suggested Tim relieve Bob of the chainsaw as he had been cutting for four hours, plus I added, he is several years older than Tim.

Bad mistake! To make a good trail you keep it as straight and level as possible. Tim would go around an obstruction instead of cutting an opening through a downed tree and Eva who was flagging the trail didn't take into consideration the type of terrain the three wheelers could safely travel over, therefore parts of the trail were on a side hill. Bob had a difficult time bringing up the machines and managed to tip each one over in the same place.

"Not good," he grimaced as he uprighted Big Red.

We noticed the plywood bottom of the trailer was coming apart for the second time caused by the heavy barrel bouncing and rocking over the rough trail. We used the come-a-long to hold it together and hoped it would make it to where we made camp that night. As we reached a flat area at the top of the ridge around 9:00 p.m. Bob exclaimed, "This is where we make Camp #2!"

We all were exhausted after twelve hours on the trail today, and agreed. After a bite to eat, we wearily crawled into our sleeping bags at 9:45 p.m. although it was still light. Another big day tomorrow!

Figure 2: Camp #2.

Searching For The Trail

The morning of June 18th we left camp at 9:00 a.m. in search of the trail Eva said would take us to Ben Leclerc's trapping cabin. Finding no trail we continued up the next ridge crossing another burn area on the way. While the three of us walked, Tim insisted he drive the four wheeler and managed to plow through the brush up the hill. I was carrying the lunches in my backpack but each carried his own drinks just in case we separated, as one needs to take in lots of fluids. It was hot! We were all thirsty but had to ration the drinks we had; therefore we were also looking for a source of water.

We checked for old blaze marks on spruce trees or a sign of manmade cuttings. Nothing! Eva had been in the area fifteen years ago on horseback accompanied by her two young sons at that time, but with the burned forest and brush growing, things were changed.

We had to descend a steep hill so Tim left the machine on the ridge to follow us. We soon realized we were on the wrong side of the ridge. A few minutes after Tim returned to the machine, Eva shouted, "I hear water! Come, Sandy," she called. "Come for a drink!"

It was a small trickle seeping out of the hillside but we managed to fill the two containers we had and Sandy drank to his heart's content. In fact, we all did! With a renewed spurt of energy we climbed up the hill where we met Tim on the machine. Eva handed him the canteen full of water.

Bob checked the topographical map as we made our way around the ridge and said to Eva, "This is approximately where one of your prospect sites begins."

She put a prospecting site marker, on a nearby tree. Then she rode on the machine with Tim. Several years ago she had been hurt in a snowmachine accident that left her with a bad knee. But first, she handed Bob her .45 pistol.

"Most likely you will lag behind and may need it for protection," she said.

As we came around the last ridge as we headed back to camp, we found ourselves looking down the valley of Koda Creek. The creek appeared to be a long way down, wet and brushy. Maybe we would find water where the map indicated one of the feeder creeks that flowed into the creek and, if Eva remembered right, the water that flowed by the cabin was similar to Squaw Creek.

We arrived at camp before them even though we had walked. Upon their return, Tim told us he had gone over on the machine when he tried turning around on a steep side hill and was pinned under it. Eva managed to lift the heavy machine high enough for him to pull himself out.

"This is not good," I whispered to Bob. "Suppose he had gotten hurt."

We ate the lunch I had carried all morning, hung our tents, sleeping bags, etc. high in the burned trees, rolled the supply barrel in the shade of a bush to protect it from the hot sun and left camp at 1:30 p.m. It took us one hour and fifteen minutes to reach Camp #1 and just think, it took us eight hours to clear trail to Camp #2.

Heading Back

We made sure everything was intact at Camp #1 before continuing down the trail at 3:10 p.m. with four and a half-hours back to the boat.

It is difficult to describe the trail we must travel over. It is no more than a deep rut which was most likely an animal trail, that has tree roots crisscrossing it which caused some deep holes between the roots. Most of the time we managed to keep the front wheel to the side of the rut but it was not easy. Actually, at times impossible! It was a really rough trail in some sections.

Tim reluctantly took my machine down Squaw Peak for the second time. I went ahead so I could video-tape them on the way down and found it really was a lot steeper than I thought!

Bob came down first, had some difficulty in one section and was followed by Tim. Again, Tim had a problem in the same spot that is sloped the wrong way on the side hill. This caused the machine to lean downhill and with the controls, like the throttle and brakes on the down hillside, he had to lean away from them. Then, he had to take his foot off the brake pedal and let go of the hand brake, therefore the machine would pick up speed. Tim almost lost control of it as he tried to keep the machine upright but managed to get his foot on the brake pedal and grab the hand brakes. It sure scared me for a minute when he seemed to have lost control.

I yelled at him, "All the brakes! Use all the brakes!" What a relief when I knew he had regained control again. Deep down I was glad it was him on the machine and not me.

Tired, lame, hot and thirsty, including Sandy who had walked the whole distance, I asked Tim if he would take my machine over the rough by-pass trail around the trappers cabin, but he refused. He said he too was lame, but volunteered to hold the back of the machine for balance and support, which did help a lot.

Now on the good part of the trail, we stopped at Squaw Creek with the rickety plywood bridge where Sandy swam and drank while we took a ten-minute break. We saw three moose near Rion Lake and Sandy swam out to greet a beaver that was swimming nearby but, the beaver went under eluding the dog.

Within a mile of Fish Lake, still on the trail, we noticed it was cooler with quite a change from the heat of a few miles back at Squaw Creek. Tim traveled around the lake while we loaded the boat with the two three wheelers, trailer, extra gear, Sandy and the three of us. Crossing the lake in a stiff breeze, it began to rain as we approached the boat ramp.

Eva commented, "At least we can be thankful we have made it all the way back without getting wet."

Satisfied we had made another safe trip with lots of supplies, we arrived home at 7:45 p.m. exhausted and lame. Bob reminded me of a saying we had read of people in the brush or bush and it goes like this!

Sometimes it pays to have a little less sense and a little more of a stubborn streak. We returned triumphant, bold but still brainless.

"Wasn't it more determination than brainlessness?" I asked.

"Maybe," he answered.

Those of us who live in the "brush" are within a road system but live twenty-five or more miles from a town. Those who live in the "bush" are away from any road system and must hike, fly or travel by boat to reach their property.

CHAPTER TWO

Leaving as Planned

We planned to leave July first for four weeks of prospecting and searching for the lost gold mine but Tim and Eva could not leave until the end of the week. She planned to bring two horses. Star would be the packhorse and she would ride Chipper. Tim would follow on the four wheeler.

Determined to leave as planned, we decided to go by ourselves. We left the morning of July 2nd. We had asked the Park Ranger at Fish Lake, the week before, for permission to park the truck and trailer at the park for a few weeks. He had us park in the overflow area for it was the Fourth of July weekend. There is excellent fishing in the lake and the Delta area is known as the playground of those who live in the Fairbanks area, one hundred miles to the north.

My Fears

For days, once we decided to go alone, I had these awful thoughts of "What-ifs" running through my mind.

What-if? Something happened to Bob.

What-if? Something happened to me.

What-if? I had to drive Big Red back if he got hurt.

What-if? I got hurt, could he handle the emergency.

What-if? We encountered bears.

I silently had to convince myself that we both had the skills but then wondered.

What-if I panicked?

What-if Bob did?

Twenty-five miles from help, over a rough trail, in an emergency, would feel like twenty five hundred miles. I wondered if Bob felt the same but decided not to let him know my fears.

Our Fears

Once on the trail we both admitted we were nervous traveling so far out by ourselves. I remembered the saying he had mentioned after our last trip.

I said, "Maybe we are brainless."

"Oh no, its the challenge that makes us determined to go. Maybe the gold," he quickly answered.

We had to travel on the trail around the lake and as I had stated earlier IT WAS NOT A GOOD TRAIL! We took time to clear detours around the more difficult areas, as we would have to return over the same trail. I really felt nervous and admitted it to Bob.

"My legs shake every time we stop and I get off the machine," I said.

He looked at me with that smirk of his and said, "Oh, you too! I thought I was the only one. Lets take it real slow and easy and take no chances of tipping the machines over."

He knew he would not make it up the one steep hill with the heavily loaded trailer so we first carried most of its contents up the hill and then, he barely made it with Big Red and the mostly empty trailer. He offered to bring my machine up, and had no problem, even though it too was heavily load.

We joined the main trail, one and half-hours later, at the backside of the lake. We then began to relax, for we knew what was ahead and were more confident we wouldn't or shouldn't have many problems, if any.

There was more water and deeper mud holes through the boggy section of the trail caused by the recent rains of late June, but we made it through with no problems.

Bob took two partial loads up Squaw Peak and on the second trip, I followed him on my machine. Let me explain something. I don't mind driving the machine uphill, but there is something about descending a hill or slope, on a three wheeler especially when the trail tilts slightly in some sections towards the downhill side. I feel as though I have no control of the machine and that's one reason why I asked Tim to take it down on the supply trips.

CB

I had called "Breaker, One Nine." to check the CB radio, while resting at the top of Squaw Peak. Remember, I said, if there was a way to hook up the CB, Bob with his Yankee ingenuity would figure something out? Well, as it turned out, most ATV's have an accessory battery powered outlet, connected to the battery system, which can be used to power a light, etc. so he adapted the CB wire to plug into the outlet.

I tried several times to contact someone and the person who did answer was not coming in very clear. It took several minutes before we could understand what he was saying. He slowly repeated, "What do you want or whom are you trying to call?"

I explained the situation to him and he was surprised I was transmitting so far! He was located at Alyeska Pump Station Nine, approximately twenty miles to the south as the crow flies from where we stood. This was the only contact we had made with anyone on the CB the whole time we were out prospecting. We could receive at Camp #2 but not transmit.

Alyeska has twelve pump stations located along the oil pipeline on its eight hundred miles of the pipeline from Prudhoe Bay to Valdez, with only ten ever put into operation. Recently, several were closed when the oil supply from Prudhoe Bay declined to less than a million and a half barrels per day.

A Looong Day

Everything was intact at Camp #1. We gathered needed supplies in the barrels before continuing on to Camp #2. With more weight added to the machines, we carefully worked our way over the rough trail. Some sections were difficult to traverse while on the machines, so to be safe, we walked them through.

I was relieved when we arrived safely at Camp #2. It had been the longest seven and half-hours I had ever traveled! Again, the fear nagged me, as I thought to myself, was it possible we were twenty-five miles from civilization and by ourselves!

We were both psyched and couldn't seem to settle down once we arrived at camp. Bob checked the topographical map while I made supper.

"The map indicates a possible water source, about a mile from camp," he said. "Do you feel like checking it out?"

As tired as I was I just couldn't settle down, so I replied, "I'll go if it means we may find water."

We hiked in the direction indicated on the map and descended a steep hill. We could hear the faint sound of water running below the ridge we stood on and as we continued further down we realized we were looking down upon treetops in the valley below us. Disappointed, the water was further than we thought. We returned to camp.

Settling In

"Let's set up camp and in the morning we'll look for water. We have enough with us for a day or two," Bob said when we returned to camp.

We took the camping gear down that had been hung in the fire killed trees and found everything okay, except for slightly damp sleeping bags. After three inches of rain the week before it could have been worse.

While we set up camp, clouds began to move in, so we prepared for rain. Bears had not invaded Camp #2, although a few things had been slightly moved and a coffee can containing nails had been tipped over. I looked at Bob, to see his reaction to the can, but couldn't tell what he was thinking so I asked, "Was it a moose or a bear?" He hesitated a moment and answered, "Probably a moose knocked it over as he passed through camp." I thought to myself, "Oh sure, he would tell me that so I wouldn't be concerned." I still had my doubts as it could have been bear.

We took precautions and set our sleeping quarters away from the cooking area. A small tent does not offer much protection from large critters, so hoped it would be attracted to the kitchen area, if one should enter camp.

I cooked a delicious road-killed moose steak over the hot coals for supper. It was given to us a few days before our trip by our neighbors, the Brennans, who are on a list for a road killed animal.

Alaska has a list of people willing to salvage road killed animals, such as caribou, moose or bear. If you request your name be on the list and are called, you must salvage the meat as soon as possible after the call. Sometimes, having to travel up to ninety miles away just to retrieve the meat.

Once it began to rain, Bob set a five-gallon pail under a corner of the tarp hung over the kitchen area to catch rainwater that would run down the tarp and into the bucket.

Exhausted, we crawled into the tent to rest these weary old bodies. I had been sleeping soundly until Bob awakened me, around 1:00 a.m. when he whispered, "I thought I heard something nearby in the bushes."

We crawled out of the tent to check. Bob, with the rifle ready and I with the .44 Mag. revolver. We returned to the tent after awhile, satisfied all was well in the area. Then I couldn't get back to sleep!

We had purchased a .44 Black Hawk Ruger revolver from a military person stationed at Eielson Air Force Base in North Pole, Alaska, the week before the trip. It is easier to carry a revolver in its holster around the waist than the rifle with its strap slung over the shoulder.

In Search of a Water Source

After a hearty breakfast at 6:00 a.m. (I couldn't sleep) I packed a lunch, cleaned the kitchen area and made sure all the food was locked in the barrels. We were ready for our day's excursions.

I became exasperated when Bob said he had found KFAR, a Fairbanks radio station located approximately ninety miles to the north of where we stood, on the portable radio.

"I have to listen to Paul Harvey, (the news commentator) before we leave," he said.

The first feeder into Ahtel Creek was about two miles, according to our estimate. We carried several empty containers in our backpacks to fill with water, whenever we found it! We followed what appeared to be an animal trail between Koda and Ahtel Creeks, traveling some distance before we descended into the valley. We found a trickle of water and Bob stopped to pan for gold as we followed it downstream. He found magnetite, which is a black magnetic sand found in areas of gold bearing gravel, and associated with gold.

Now encouraged, as he had found the magnetite, I tried my dowsing rods, but found little response. Yes, one can dowse for gold, also!

This trickle eventually, joined a slightly larger feeder of Ahtel Creek from a different drainage. Bob used the machete to clear the way as we maneuvered our way through the thick brush and made our way upstream a short distance. I carried the rifle. We filled the containers just before it became a trickle of water seeping out of the moss.

We began to climb back up towards the ridge that we had followed that morning. Ever so slowly, we climbed the steep 45 degree angle hill covered with a thick layer of moss with rocks, hidden beneath. Bob carried three and a half gallons of water, while I carried only one, as I also was carrying essentials like first aid kit, rain gear, extra socks, the lunch, two quarts of tang flavored water that we take with us whenever we hike or prospect.

Having already traveled three miles that morning, we were tired! It was a very difficult climb at such a steep angle over the moss covered, rock strewn tundra. And cruel to our legs.

We joked about our ages and rested often. Hey, we are getting old and so are our thirty-year-old backpacks! These were canvas Boy Scout packs and purchased when we first started to backpack in NH. Which will give up first? Us? Or the packs?

As we neared the top, we found hidden amongst the moss-covered rocks, a small trickle of water. We enjoyed a cold refreshing drink with our lunch and were thankful we didn't have to use the water we were carrying. In the heat of the day, we couldn't seem to quench our thirst and had to drink small amounts of fluids often.
We arrived at camp by mid-afternoon, exhausted, dirty and sweaty after an estimated five-mile trek.

Bob said, "After all the work of getting the water, we deserve a shower!"

We limited ourselves to three quarts of precious water, between us, for our showers. When the right temperature had been reached, Bob hung the can from a branch of a secluded tree. The spigot would release a small stream of water when it was turned on and was enough to cleanse the body.

Shortly after our showers, we heard the sound of distant thunder over the Tanana Highlands and hoped the storm would not come our way as Camp #2 is atop a ridge with no protection from a thunderstorm. We were thankful when it headed away from us.

The Tanana Highlands is a ridge of 3,800 to 4,000 foothills along the northern drainage of the Tanana River and for reasons unknown, creates its own thunderstorms.

Again around 1:00 a.m., I woke with a start to the sound of branches rustling, as if something was circling camp. I decided not to wake Bob until I was sure there really was something out there.

I listened for what seemed like an eternity, my heart pounding, and all the time my hand on the revolver. I was ready! Ready for what! I didn't know. I wondered, "Was it a bear? Or a moose?" Nothing? I heard nothing more but needless to say I slept, if at all, very lightly after that.

Ridge Trail

We knew it was going to be a hot day as early as 6:00 a.m. with the twenty-four hours of daylight. We planned to spend the day cutting trail along the ridge we followed yesterday and left camp early, even before Paul Harvey News!

Throughout the morning, Bob cut with the chain saw and I cleared the brush. Every one thousand feet or so, I would return for Big Red. Although I carried the revolver, I didn't tell Bob I was frightened to walk back over the trail, but I knew I was the one who had to do it.

I would scan the brush as I hurried along the trail, fearful of a bear encounter. And you know what! The rifle was in the case ON the machine!

The morning heat was bearable with a slight breeze to cool our sweaty bodies so we continued to work until noon. Then, while we rested, we wondered if the trail was the old Indian trail between Forty-Mile country and Fairbanks or an animal trail? Could it be the horse trail Kosloski used while guarding the flats from hunters, trappers and miners. Maybe it was the area where he found the gold.

The ride back to camp with both of us on the machine proved to be rough, but it was much easier than walking.

While we laid in the tent to rest after lunch, Bob said, "Let's hike back down the trail to find where we missed the trail on the way into Camp #2 on our second supply trip."

I lay there, sighed, hesitated, but then agreed. We did need to find a water source closer to camp.

We backtracked, on foot, about three quarters of a mile,

determined to find the old trail that we had missed and found where it veered to the right along the ridge of the heavily burned brush, which hid it from view. It was not easy to follow, but Bob could see the potential of us cutting a trail along the ridge. According to the map, we would come out onto the trail we had used on our first supply trip, at the headwaters of Koda Creek. Lo and behold, we emerged out of the wooded ridge and onto the same trail. We continued to follow it part way up the next ridge to check if, with a little work we would be able to drive the three wheelers up one steep section. We decided it would require a lot of work.

The Water Source

As tired as we were, but desperate to find a water source for camp, we veered off the trail to check the area where the topographical map indicated Koda Creek began. The terrain was a lot steeper than we had anticipated as we worked our way down through the dense thicket. Bob used the machete to cut a way through. When, about 500 feet from the trail, we heard the faint sound of running water. We traveled slowly through the brush and came upon a small spring seeping out of the hillside.

I yelled, "Yippee! We found some."

Unquestionably this was now our water source! We filled the container we had with us, then returned to camp, a distance of a half-mile. As we climbed up the slope to camp, I noted that the grass on the side hill would be feed for the horses when they arrived.

While I prepared supper, Bob said, "We can take a shower every night!"

Shortly thereafter, a cool wind began to blow lightly so we hurriedly took our showers before it began to rain. As we crawled into bed, I made the comment, "Oh yeah! We got to clean and rest these old bodies so they will be ready to go again tomorrow."

Once more, around the same time, something circled camp. Now the adrenaline was working as we listened. Ever so slowly it moved through the brush. We both held our guns, looked at each other and silently wondered if it was going to enter camp. Then we heard it move away. What was it?

The Mountain

The morning of July 5th was gorgeous! We planned to find the trail Eva said should lead to Ben Leclerc's cabin. While Bob listened to Paul Harvey on the radio I packed the lunch and two quarts of tang flavored water, then put the rest of the food into the barrels. Yes, I learned to be ready after Paul Harvey.

We followed the same trail as on the last supply trip with Tim and Eva and found ourselves on the left side of the ridge. Again, the trail would disappear, then reappear as we followed it through the thickets. After some time, we emerged out in the open to a treeless arctic tundra, strewn with boulders that were covered with thick, soft moss.

Determined that this was the right trail, we continued on to the next ridge to prospect in the valley of a feeder creek. This proved to be more difficult than we had anticipated, as we had to be cautious at all times. It was dangerous, with holes as deep as four feet or more, hidden by moss, next to the boulders.

Figure 3: Florine crossing rock slide.

Part way across the tundra we traveled through a section of thick alder brush, which is an indication of a wet area. It too, proved difficult with having to maneuver around fairly deep water holes in the low, thicket brush. We had to be alert to the possibility of a fox, wolf or bear which could be using this area for its protection. To warn them of our approach, we carried on a "useless" conversation.

Once through the thick alders, we headed toward a rock slide at the further end of the ridge. It was difficult to work our way across the steep rock slide with its large, haphazardly strewn boulders. We could see the slide ended several hundred feet below us in the creek where we had planned to prospect.

We edged our way down to where the creek should be, but found it flowed under the rocks. Bob managed to get a cupful of water at a time by lying on his stomach and reaching down between the rocks. He noticed ice wedged between the rocks and pried off a piece. Oh sure, on a cool day we find ice for our drinks.

This reminded me of a late August backpacking trip, several years ago, in Gorham, NH when hiking the Icy Gulch trail. We discovered ice hidden beneath the huge rocks of what is called Icy Gulch.

"Let's go see what is on top!" Bob said enthusiastically.

"Do you think you can make it?"

Without hesitating I replied, "If you can, I can! Let's go!"

For the first two hundred feet we slowly climbed upward, stepping from one rock to another like climbing a steep stairway, as we ascended at a 45-degree angle of another rock slide. All of a sudden, a Pica appeared on a rock a few feet ahead of us and seemed quite disturbed to have strange intruders in his territory.

I whispered, "I wonder if he has ever seen a human before?" After a moment, Bob quietly answered, "Probably not, as it could be a young one."

A "Pica" is a member of the rodent family and is like a large mouse, but lives among the rocks at higher elevations.

As I glanced back to look at the view, I noticed several Gray Jays had begun to follow us and thought it strange that they were at such a high elevation. A Gray Jay is also known as a Canada Jay and is called a Whiskey Jack in Maine. It has also been called a Camp Robber for it can be a pest at a campsite stealing food off the picnic table.

As we continued to climb higher, we heard a shrill whistle as we meandered through open tundra near the top. Bob immediately spotted the hoary marmot as it sat on a rock, a short distance from where we stood. It blended so well with its surroundings, that it was difficult for me to spot it.

These creatures are relatives of the woodchuck. They live among rock piles, having several entrances to their den beneath the rocks. They are also found at lower elevations on rock ledges above creeks. This one sure was curious as to what we were, and it too, had apparently never seen a human.

The Top

After four and a half exhausting hours from camp, we crested the last ridge and were at the top! With a renewed spurt of energy we crossed the half-mile mountain top. Why? To see what was on the other side!

The top was a rock-strewn plateau carpeted with wild alpine flowers in full bloom. A forest fire burning in Yukon Territory, Canada, many miles to the east, left a smoky haze in the air but the view was still spectacular!

We looked down from our vantage point (from the other side) into a valley where sections of a river flickered brightly in the sunlight as it meandered towards the Tanana River.

To the south, Donnelly Dome stood out by itself with the tall mountains of the Alaska Range in the background. The farmlands in Delta Junction spread out for many miles beyond the large river. Mostly barley and potatoes are raised on the area farmlands.

The Delta River was visible in the distance where it joined the Tanana River, beyond the bridge on the Richardson Highway. The Tanana River is the only place the eight hundred-mile oil pipeline is suspended across a river.

We came upon a pyramid shaped metal building located on the highest point of the mountain. It had two antennae protruding from its top and was powered by two solar panels. It was the forest service transmitter Eva had told us about. Oh yes, man had been there! We found miscellaneous junk such as rusted cans, wire, a barrel, etc. where the fire tower once stood when it had been manned to spot forest fires, back in the 1950s.

We heard the shrill whistle of another hoary marmot as we

crossed the plateau, warning others of strange intruders in their territory. It sat atop a pile of rocks known as a cairn. It quickly vanished amongst the rocks as we drew closer, reappearing a short distance away. We stood quietly observing it when another whistled from the rocks below us. It appeared to be a young one. Now alert to our presence, it nervously watched us. With a mouthful of grasses it vanished beneath the rocks only to reappear several feet away. Winter begins as early as mid-September at high elevations and they apparently were preparing for the long winter

Bob discovered, while picking blueberries, a brass marker within twenty feet of the cairn embedded in a rock. A cairn is a manmade marker and usually a pile of rocks to mark the trail or as that one did, a survey marker. The brass marker had been put there by the Geological Survey, stating that for more information the Survey Department Washington, D.C. could be contacted, it was marked with a number one and an arrow pointing towards the cairn. I then noticed, several feet to the east, another one with number two embedded in it and an arrow pointing towards the cairn. The rock cairn was a survey marker, which represented a bearing tree. The brass markers represented the three surrounding trees of a survey marker.

We climbed the mountain that day because the temperature was cooler but as we dropped off to the southwest, it became quite warm. Coming down was much easier on the heart and lungs but at times difficult on the knees. Guess you can't win one way or the other when you climb mountains at our age.

On the first plateau, below the mountain, we noted someone or something had recently overturned rocks. I reminded Bob that the author, Paul Solka, told us his nephew planned to fly to the area in early May, via helicopter. Bob wondered if he had landed and overturned the rocks. Me? I wondered if it had been a hungry bear!

Bob thought he would find the trail that we had come up that morning but missed it somehow. With no trail, we plodded through the thick brush and made lots of noise even though we had seen only one moose, off in the distance, that day, But what circles camp at night? I wondered.

We emerged from the thicket onto the helicopter pad we had found on our June trip with Tim and Eva. We followed the flagged trail we had marked at that time and returned to the three wheeler left by the trial.

We filled a few containers at the water source and carried them up the steep hill to Big Red. Carrying the containers back to camp was much easier with the machine. Unfortunately, one of us had to walk. Guess who?

"Only one of us should ride because of the extra weight of the water," Bob said.

He knew I didn't like to drive Big Red as it is a much bigger machine than mine. He handed me the .44 revolver and said, "I'll take your pack. You take the revolver and while I finish filling the last container, you start back to camp."

He caught up to me when I was half way up the steep hill to camp. He stopped and said, "Get on, I know you are tired."

I was surprised at the offer, but didn't hesitate very long and as I got on I slyly remarked, "But you said only one should ride."

He glared at me. It wasn't long when the front wheel came a foot or so off the ground as he drove over a tree root, but he managed to keep the machine from going over backwards. I thought to myself, I should get off! My mind said I should, but my body said, oh no, too tired!

Day of Rest

The morning of July 6th we were awakened with a start at 6:00 a.m. We realized it was the sound of a moose running towards camp. We looked at each other, wondering if it would turn away before camp. Still looking at each other, half asleep, we let out a sigh of relief as it suddenly stopped. Why?

We heard the brush rustle behind the tent so I grabbed the camcorder and got a video of the moose through the small rear window of the tent. She stood still about forty feet away and looked our way. She must have decided we were of no threat to her as she turned and trotted off. Boy, it sure got our old hearts abeating! It's scary sleeping in a tent with no real protection around you. Try it sometime, especially with wild animals wandering around.

Our bodies needed a rest after our long climb up the mountain yesterday. We set up a more comfortable campsite. Bob cut poles and used the plywood to erect a work area for preparing food and washing dishes. He wanted me to be comfortable while I prepared HIS meals!

We enlarged the tarp over the kitchen area by adding another, which sloped down to catch rainwater in the half barrel used for panning samples. We overturned the trailer for a table and used milk crates for seats. And the latrine "Dunny" now had a comfortable seat where we could enjoy the birds and surrounding scenery. Ah, peace and quiet.

With a comfortable camp setup and the lunch dishes done, I had settled down for a quiet afternoon of "rest" when Bob announced he was bored., "Let's cut a trail to the water source," he said.

I contemplated what he said and thought to myself, I really wanted to rest but don't want to stay in camp by myself. I let out a long sigh and looked at him.

"I suppose it would limber up my aching muscles. Besides I need to wash a few clothes," I said.

We both rode on the machine to where we would begin to cut the trail. We cleared five hundred feet or so of trail to within one hundred feet of the water, When it became too steep for the machine, we continued to cut a narrow path the rest of the way to the water source.

Bob used the small can, left behind for such a purpose, to scoop water into the containers. He filled two five gallon buckets for me to wash and rinse a few things plus what we needed to carry back to camp.

When I finished the wash, we carried the water containers and clean wash up the steep hill to the machine. Bob went ahead of me with the water and wash while I walked back to camp. It was hot and boy, were my legs tired! So much for "Our day of rest!"

Soil Samples

As usual, we got carried away and found ourselves further down in the valley than anticipated and climbed back up the steep hill to the three wheeler. But oh, what a blessing to ride back to camp in only twenty minutes instead of the long hike it would have been.

We had taken a class on Geo-Chemical testing at the University of Alaska Fairbanks (UAF). Samples are gathered by using a hand auger to bring soil up from a depth of three or more feet. The samples are put into a container and marked with a number. To locate where the sample had been taken the same number is written on a strip of bright colored flagging tape, then tied to a nearby tree. Next, the number is written on a hand drawn map showing its location.

The samples are dried and screened through nylon to get only the very fine particles. A small amount is put into a glass test tube. Then a solution of combined Dithizone, Coleman fuel, salt and distilled water is added. The tube is shaken for one minute. If minerals are present, the solution will turn red, blue, yellow, etc. depending on which minerals are present but this does not indicate the presence of gold.

This is what is called Geo-Chem Processing and is time consuming, as it took me two hours to test fifteen samples. I was disappointed when they tested only slightly mineralized.

Bob went after water while I stayed in camp working on the samples. That was the first time I stayed alone and, of course, I became aware of the surrounding brush and sounds of nature. A branch would creak. A leaf would rustle in the breeze. The wind whispered, birds sang, and the sudden chatter of a squirrel nearby startled me. Don't worry, I had the gun by my side.

It sure made a difference with a trail cut closer to the water source as Bob returned within fifteen minutes, but it seemed a lot longer to me! I thought that maybe he didn't like to leave me by myself and had hurried back. If he had, he wouldn't admit it.

They Arrived

On the seventh day, with a thick fog and drizzle, we stayed in camp and kept warm by the fire. We listened to a Fairbanks radio station, KIAK, located ninety miles to the north. Bob commented that you still need outside contact with the world even if you live in a remote area.

While preparing dinner we heard the sound of bells jingling and Eva's voice through the thick fog. I quickly added more food to the pot and several minutes later they appeared out of the fog with Sandy leading. Eva rode Star, the younger horse, while thirty-year-old Chipper was used as the packhorse. Tim followed on the four wheeler. They had been on the trail for two days, camping at Rion Lake the first night, then at Camp #1. They could have made it all the way into Camp #2 but they weren't sure if we had moved or not. It was nice to have company but it had been a glorious week by ourselves. Like a second honeymoon!

Eva tied the horses to trees in the grassy area after they had taken them for a drink and filled several containers. It was damp and cool so we kept a large campfire going to keep warm. It was nice to have company for a change, but, as usual, we crawled into the tent early to keep warm. Sandy slept in their tent because the mosquitoes were miserable during the dusk of the night. Eva rubbed mosquito repellent, used especially for horses, onto each horse and tied them close to camp. Oh, I thought, I can sleep tonight! Let someone else listen for that animal that circles camp every night.

The Cabin

While watering the horses the day before, Eva noticed old blaze marks on a couple of trees below the trail we had cut. She thought that might be the trail to Ben Leclerc's trapping cabin, so we set off in search of it.

As usual, we had our backpacks and carried the lunches, drinks, soil-sampling supplies, first aid kit, rain gear and gold pan. Bob carried the shovel and machete. The machete, which was used to cut small branches and brush blocking our way while hiking. Most of the trails we cut are with the chainsaw. Eva followed, leading the horses and Tim, who thought he could drive the four wheeler, soon realized it was impossible. He also forgot to bring his gold pan and shovel.

We found several more old blazes part way down the trail but they soon disappeared. For over two hours we hiked steadily along the brushy trail before we stopped to rest. As a typical routine, Bob would disappear with gold pan in hand, only to return when he thought lunch was ready. Upon his return, he stated that he had found only pyrite in the creek. Pyrite is known as fool's gold. Tim, who had gone ahead of us, also returned in time for lunch.

Over lunch, there was a great deal of discussion of where the cabin might be. Eva said, "The area and trail look so different than how I remember it fifteen years ago when I came out here with my two sons on horseback."

After lunch Tim and Eva decided to return to camp. Not us. With all the hiking and exploring we had done in the past week, we were ready to continue, curious as to what was beyond the ridge on the other side of Koda Creek.

To cross the creek, which is a narrow trench carved into the perma-frost ground and fairly deep in some areas, proved rather difficult. The brush along the creek was thick and low and we had to work our way downstream before we found a narrow enough place to cross. Once over the ridge, we had seen from the other side, we continued over another to check a second feeder creek. This creek was also a trickle of water with deep holes.

Bob panned, but found only a small amount of magnetite in the gravel he had managed to dig out of the creek. Remember, magnetite is a magnetic iron ore, black in color and usually an indication of the possibility of gold in the area.

Checking the map as to where we were, Bob noticed that this feeder creek was part of my prospecting site, Ma VI, so we tagged a tree. I suggested we continue following Koda Creek down further as long as we had gone that far.

Bob quickly answered, "Oh, no. We've come far enough for today. It's a long hike back up the valley to camp."

"Oh, all right, " I responded, knowing he was right.

On the way back, we found a fairly decent trail on that side of the creek and wondered if it was an animal trail or manmade. The question was answered when the trail passed through a downed tree trunk that had been cut many years ago with some type of saw, an indicator someone had worked on the trail. It may very well lead to the trapping cabin Eva wanted to find.

Some sections of the trail were rather difficult to traverse because of the many recently toppled trees lying across the trail. Whenever the trail skirted close to the creek, Bob would pan but found nothing. Continuing, we found a cabin hidden along the trail.

Bob said, "This must be Ben Leclerc's trapping cabin!"
The cabin was fairly solid and low slung. It had trees, three and four inches in diameter, growing in the thick sod covering the roof.

Figure 4: Remote trapper's cabin (note trees growing on sod roof).

Apparently, no one had been there for sometime to maintain it, as the few doghouses nearby were slowly settling into the earth. What a mess! Bears had ripped the inside of the cabin apart, most likely searching for grease or salt that may have been left. Once a bear enters a building it never goes out the way it came in therefore, creating a lot of damage.

Curious as to the location of the cabin we checked the topographical map and noted the creek that flowed nearby was the same feeder creek with the rock slide we had crossed close to the summit of Pedro Dome last week.

It appeared the trail ended at the cabin and we knew Tim and Eva would want to see the cabin so we marked the trail as we crossed over Koda Creek to the trail we had come down that morning.

A Late Night Visitor

I awoke to the pitter-patter of raindrops falling on the tent. Or was it something else? The time was 1:30 a.m. the usual time of our nightly ritual of hearing something circling the camp, only this time it was a crunching sound I heard. I anxiously listened, aware that the adrenaline was beginning, for the sound was close by. Was it the horses? Sandy? After several minutes, which seemed like an eternity, I heard

Tim and Eva's tent being unzipped. Oh, must be the horses I thought as I rolled over.

All of a sudden "Kaboom!" Bob sat bolt upright out of a sound sleep.

He loudly questioned, "What's going on?

I too wondered, but then Tim answered, "A bear. Didn't you hear it?"

We quickly emerged from the tent, guns in hand and rushed to where Tim stood just twenty feet behind their tent. Only a few feet ahead of him lay a medium size black bear. It stirred and let out a mournful cry, which sent shivers down my spine, and then it was silent. In the meantime, Eva joined us. She was laughing. .

"There was no way Sandy would come out of the tent. Our watchdog!" she said.

She thought it strange the horses hadn't warned us, but maybe they hear, see or smell bears when in the corral at home and never were bothered by one. So they have no fear.

Tim used the four wheeler to drag the bear away from camp. While they began to skin it, I started a fire to make coffee. I knew it was going to be a long night! Sandy was at my side when he let out a quiet "woof" as he faced down the trail. I listened intently, and then heard branches rustle a short distance away. Was it another bear? Was it Mama bear? I hurried back to where they were dressing the bear and had the gun, in case whatever it was, decided to come into camp.

Eva stated, as she skinned one front leg, "Once a bear has found food in camp it'll become a pest, always returning for more. You have to get rid of it before it becomes one."

We realized there was no way we would use all the meat before it spoiled so kept the hind quarters only and buried the rest with the hide. Under normal conditions you would salvage all usable parts, including the hide. At 4:00 a.m., the early morning temperature was a cool 50 degrees. Bob hung the small chunks of meat in trees to cool and we crawled back into our tents.

Bob got up early to take the meat down before the sun and heat of the day would warm it. We spent the day cutting and cooking the meat. To cut the meat a cutting board was needed so Bob headed for the woods, chainsaw in hand, returning with an 8 x 14- x 2-inch plank he had carved from a downed spruce tree. He and

Tim took turns cutting the meat into steaks and stew meat.

The stew meat was combined with garlic, onion, salt and pepper and boiled for four hours before it was tender. To store it we put it into large coffee cans and put these in the ice cold water of the water source.

The steaks we barbecued for lunch were chewy, but tasty, so I marinated the rest in hopes it would tenderize them. To keep these cold, Eva and I dug a three foot hole in the ground and set a five gallon bucket inside. We then covered the hole by laying logs across the top, spread a piece of plastic over them and shoveled dirt on top of the plastic. It worked great and the fresh meat kept very well. Actually, ten days later, we brought some home with us.

The next day it rained so we were forced to spend another day in camp. It remained a cool 54 degrees all day, although there were twenty hours and twenty-six minutes of daylight at that time. It was a long restless day. We read, Tim and I played cards and by evening Bob was restless. He picked up his chainsaw to start it. It was getting harder to start and at times I was afraid he would have a heart attack from pulling the rope so many times. He didn't dare take it apart to fix it as we couldn't very easily run to the store for parts.

Once it started, he cut some fire-killed trees and cut them into firewood. Apparently, it wasn't the firewood he was after but was clearing an area for a better view of Pedro Dome. Of course it was nice to have more firewood.

A Day of Prospecting

Morning of July 12th, during the early morning hours I heard the horses, as if something had disturbed them but then it was silent again. Eva said Star had been caught in his rope and needed to be untangled and so since she was up, she started the fire and had the coffee ready.

We left camp early, each carrying our own lunch. We planned to travel on Big Red to the end of our trail between Ahtel and Koda Creeks, then hike to a further feeder we hadn't checked yet. Eva followed on Star, leading Chipper, and Tim was on the four wheeler. Several times we had to wait for him, as he had to find his own trail from where ours ended. We munched on the numerous blueberries and raspberries while waiting and noted several fresh bear droppings near where we were picking the berries.

After a while we veered off the trail and headed to a helicopter pad on a ridge that jutted out towards Koda Creek. Looking down the steep slope to the feeder creek below where we wanted to take soil samples, it seemed to be the steepest of all the feeder creeks we had already checked. Eva couldn't follow with the horses so she would work her way around the ridge where we would eventually end up, once we checked the next valley.

Tim stood there, looked down the steep hill to the valley below and said, "No way. I'm not in shape for that!" We all looked at him. He looked at Eva and said, "I'll go!" After all, I thought, the agreement between us was the four of us work together. I called to him to bring his shovel and gold pan as we descended cautiously towards the valley below. Although we were camped in an old burn area which covered most of the ridges, each valley or ridge had different vegetation from thick brush to tall spruce trees, poplar groves or alpine grasses with wild flowers and berry bushes.

A short time later we noticed Tim wasn't following us which is what we expected. We took six samples as we followed the creek towards Koda Creek before stopping for lunch and a well-needed rest. I was amazed at how the birds would find us, land on nearby branches and quizzically watch us. I talked softly to them and offered bread crumbs but they were cautious and kept their distance. I wondered if they had ever seen a human before. Probably not.

The Next Ridge

With renewed energy, we continued up the next ridge where the terrain flattened out under a grove of birch trees. Bob checked the map and decided that this ridge was the northeast corner of one of his prospecting sites. We posted the proper papers on a tree and marked the map as to its location.

As we worked our way upstream through the thick brush in the second valley, we collected more samples and panned gravel in several areas. Still no black sand (magnetite). We were alert to our surroundings, as there were lots of fresh bear signs along the feeder creek we were following. The cans I had tied to my pack would jangle.

They were not as loud as we would have liked, so Bob created more noise by striking the shovel against the machete he carried.

We emerged from the thicket into a grass covered clearing and could see the top of the ridge where we expected them to meet us. Lo and behold, as we crested the ridge we found we were only half way to the top, which loomed high above us.

With no sign of them at the top, we returned to Big Red after six hours of hiking, and headed for camp where we found a note. It stated they were to spend the night at the cabin (Ben's) on Koda Creek. Ah, I thought to myself, peace and quiet, but too quiet as there was no one else but me to listen for bears!

I was paranoid after the bear episode, and with no extra security in camp I slept fretfully. At 1:00 a.m., I insisted we get up to check what the noise was I thought I had heard. Suddenly the cans we had hung over the dog food bumped together and clanged. Then, in anticipation of an encounter with a bear, we readied our guns. Both of us stood in front of the tent, hearts thumping, (at least mine was) and glanced around. Nothing! Poor Bob, either he had patience with me or he was happy someone was on watch and he could sleep.

Alone Again

In the morning, we returned to Camp #1 for more supplies and another storage barrel for Camp #2. Tim and Eva returned from their overnight adventure to the cabin shortly after we arrived back at camp. Bob asked if they had done much prospecting.

"No, it's too difficult with all the thicket along the creek but I did find a speck of gold in the creek near the cabin," Tim answered.

Bob told me later, he thought Tim had told us he had found a speck of gold so we would prospect on their sites. Well, we had our own areas to prospect. They left in the early afternoon and planned to explore the areas around Canyon Creek near Camp #1.

I finished testing the soil samples we had collected and with our bodies fully rested, we were ready to hike again. The four-mile hike took us down a valley to the main feeder to Ahtel Creek to where another joined it. There were lots of raspberries and BEAR sign as we descended down the grassy valley. Along the way we found a large moose antler in perfect condition but neither of us wanted to carry the extra weight. Besides, it provided calcium for small creatures of the forest.

That evening, discouraged with the results of the soil samples, we decided to return home. That night, between nightmares of bears, I also had a vision of finding the gold mine. In this vision, I saw a dome.

The first thing in the morning, I told Bob about my dream of a dome and my vision of finding the gold mine. I believed the next ridge from the end of our trail was Ahtel Dome as it was shaped like a dome at 2293 feet. I then asked if we could check one more area before we left. He agreed.

We traveled on foot to the north side of the dome and wherever my dowsing rods indicated (crossed) we managed to get soil samples from five feet down, using a shovel and hand auger. We noted the soil was rusty colored gravel, which could contain iron and may have caused the rods to cross.

We took several samples before continuing down off the dome at a 45-degree angle, traveling south-southwest until we came upon a trickle of water. This was the very beginning of another feeder to Ahtel Creek. Bob panned in several places as we followed it downstream but found nothing and I had no strong signal from the dowsing rods either.

We climbed a moss-covered ridge and crossed through a grove of black spruce onto the crest of another ridge. While stopping to check the topographical map, we noted that this drainage area was not indicated on the map. Proceeding a little further, we came across a small seepage of water that had a coating of rusty scum, an indication it was stagnate with little drainage.

We skirted the drainage area from the ridge above but found it difficult to walk through the thick brush, so we opted to go a little higher. Here we came upon a quartz rock, approximately five foot square! In all our prospecting in the past two weeks, we had not seen any quartz rocks. What did this mean?

I was so excited about finding this quartz rock especially where I had visions of the gold mine below the dome, until Bob said, "Don't get your hopes up. Very white quartz rock is an indication it has no gold. What we are looking for is rusty quartz. I think most of these rocks are called float rocks which means they were moved by frost or weather over many years."

We found several more quartz rocks as we continued. Some completely hidden beneath a lush moss covering. Why here? Why on this side of the ridge? Why not in other valleys we had explored? MY, or maybe I should say, OUR enthusiasm returned!

Needless to say, we returned to the area the following morning to check the location of the rocks again. We decided to post six mining claims which we named Lucky Acres I through VI.

Exhausted as we were, we measured and posted the corner of each claim. We felt this to be a promising indication that the Lost Gold Mine of the Upper Tanana River could be located in the area. It was another long day.

The morning of July 19th, wanting to continue to explore, we had to admit our bodies were totally exhausted and with the threat of rain we decided to return to civilization. Yukkkkk!

Problems on the Trail

First, Bob drove my machine down Squaw Peak, then tied a rope to the back of Big Red for me to hold onto. I stood on the uphill side of the trail while he maneuvered the machine over two tree roots protruding about six to eight inches across the trail. This caused the machine to tip towards the downhill side. I used my body weight to hold tension on the rope as he slowly inched his way over them. It worked!

Someone had worked on the four-mile trail around Fish Lake with a small dozer since we had come over it nineteen days ago. It was slow traveling over that section of the trail because of roots, trees and brush left lying upon the trail. I yelled to Bob when my machine stalled for the fifth time caused from the spark plugs fouling because of the slow travel.

"My machine stalled again!"

At the time I didn't realize he was in a predicament of his own. On a steep section where the dozer operator had left an uneven mess he couldn't stop the machine and trailer because of bad brakes. When the machine began to go over he put his foot out to keep it upright. The rear tire ran up his leg, pulled his hip boot off causing him to wrench his knee.

He had regained control just as I crested the hill calling out that my machine had stalled again. I noticed he favored one leg as he walked towards me. He explained what had happened while starting my machine, then drove it to where his was parked. We realized how lucky we were it had happened within one hundred feet of the end of the rough four mile trail around the lake.

Within a mile of the truck, it stalled again. He gave me a disgusted look, when he caught up to me, and asked, "Don't you know you have a reserve gas tank?"

Now I was upset and snapped back, "No, I don't! Besides I thought it was the plugs again."

He showed me where the switch for the reserve gas tank was, flipped it on and with an aggressive pull, started the machine for the last time on that trip. We later learned we had a bad batch of gas which is not uncommon in Alaska.

Company Coming

We planned to return to the area in late August after my brother Henry, who had a kidney transplant a few years ago, and his wife Rita left. It was their first visit to Alaska and we were pleased he could make the trip because of his health. We spent three weeks showing them our beautiful state.

In the meantime, I had received a caribou permit, which was the first in eight years. We automatically qualified for subsistence permits while living in Sourdough because we lived far enough from a large city such as Anchorage or Fairbanks. Now, we must apply for a special permit or drawing because we live closer to a large city. There is no regular hunting season for caribou and we were unsuccessful in our hunt because of little time to hunt during the short season.

Filing Claims

I tested the soil samples we had gathered in August, the week before we planned to leave on our annual trek back to NH and found them highly mineralized.

We made an extra trip to Fairbanks to file and record the six claims we had marked in August, just in case the samples proved to be mineralized. They were! Now we had something to look forward to next year.

CHAPTER THREE

Our Second Year

We were delayed in NH the first six months of 1996 due to personal business and returned to Alaska in late June. Therefore, we did not get out to our mining claims until mid-July.

Here I would like to give thanks to our neighbors, Skip Peters and Michael Fisher, for being so kind as to roto-till and plant our garden by mid-May. No words can express our appreciation for their hard work and the many meals that it provided for us. Thank You!

The first major decision to be made was whether to purchase a four wheeler or not. Money-wise it was not feasible, but safety-wise it was a necessity, as we were on our own that summer. We bought a 1996 Suzuki King Quad, which seemed to be the machine needed to haul heavy loads over the rough terrain and steep hills.

The Bridge

While visiting Tim and Eva one afternoon, they told us the trapper whose cabin we had by-passed last year had destroyed and burned the old rickety plywood bridge across Squaw Creek. Apparently he felt his territory had been invaded after he posted the PRIVATE PROPERTY NO TRESPASSING sign and we continued to by-pass his property. "We are in the process of building another bridge further up stream," he said. We volunteered to help and plans were made.

We set camp up near Rion Lake before continuing to the bridge via the newly cut by-pass trail. I had to admit that as we approached the bridge construction, it was quite impressive! Eva explained how they had to search for two tall spruce trees, fairly close together, to fell across the span of the creek to form the main beams of the bridge. They had a problem with one, she said. and had worked hard to position it parallel to the other.

Some of the bracing and cross members that would support the bridge had already been done. Bob volunteered to work on them while the three of us cut, carried and nailed into place poles of uniform size atop the tree poles (beams) that spanned the creek.

We all worked hard and diligently for several hours to complete

Figure 5: By-pass bridge.

it. Tim and Eva had the honor of christening it by crossing over first. Bob and I observed the bridge as they crossed to check if it was sturdy or not. It was, and we felt it would withstand the ice and high water during spring breakup. We crossed on our machines and continued to cut a trail to join onto the original trail above where the rickety old plywood bridge once stood.

Although the architecture of the bridge was unique it was impressive and a major feat there in the wilderness. It should be appreciated by many a traveler over the trail. Especially us!

Eva informed us she and Tim must work most of the summer and couldn't join us in our continued search for the lost gold mine.

Finally Departed

We departed July 17th. at 11:30 a.m. from Fish Lake after several weeks of preparation. Bob decided it would be safer for me to drive the four wheeler. I was excited until I realized I had to pull the heavily loaded trailer.

I traveled slowly at first, not used to the new machine or pulling a trailer. It was again heavily loaded with another fifty-gallon barrel, plus cans of gas and other miscellaneous supplies. I thought the trailer was going to tip over with its contents as the barrel rocked back and forth.

The trail was slippery from the recent rains and it was impossible to drive up some of the steep hills. We cut more by-pass trails through the brush and made it to the top without removing any supplies as we had done last year.

The Trailer

I heard a different sound coming from the trailer as I drove through the boggy area and stopped to check what was causing it. Bob thought it was okay. We followed the new trail to the ridge. It was difficult maneuvering the machine and trailer over the narrow trail and between trees. Suddenly, it was a lot easier to steer!

"Did you lose something?" I heard Bob yell. I stopped. I had lost the trailer! "I thought it was easier to steer!" I laughed.

The hitch had come apart when a bolt had fallen out. We found the missing pieces lying atop the moss a short distance back down the trail. Upon returning, Bob said, "It probably was loose when you heard the noise back a ways. Although, I have enough to repair the hitch, it is best to find the things you have lost."

Camp #1

We crossed over the new bridge, by passed the trapper's cabin and had no problems up Squaw Peak. We arrived at 5:30 p.m. at Camp #1. Oh oh, a bear or bears had been there! The yellow plastic barrel had been scratched and tipped to its side but it was still chained to the tree. The metal barrel with the locking ring was intact although the ground beneath it had been dug out by some critters.

The rugged yellow plastic barrel is the type used by the forestry service and found near wilderness helicopter pads. A fifty-gallon drum of fuel is stored in each barrel for emergencies. Eva had found that one floating in the slough of the Tanana River.

The Cache

Bob's curiosity got the best of him and he couldn't resist opening the cache barrel of the old trapper. The top section of the barrel had been cut to fit over the bottom half. It was badly rusted and difficult to pry apart, but he coaxed it and peeked inside.

What would be stored in a supply cache? Well, there were a few bullets for a rifle and an empty Prince Albert tobacco can probably kept in case of a need. The contents of the canned goods were dried because the cans had rusted through. A pound box of raisins had thirty-five cents stamped on it! A small army metal ammunition box held first aid supplies that were still usable and a 1981 penny! Was it put there by the trapper or did someone else open the cache and place the penny inside the box? We will never know!

Camp #2

We continued on to Camp #2, although we were exhausted. Bob thought it would take us an hour and a half at the most. Wrong! Clearing the blow down through the burn areas took over three hours!

We arrived, after traveling a total of nine and a half-hours, to find camp had been visited by a bear or bears. The trash was scattered about. Three of the five one-gallon cans of gas had large holes from being bitten and were empty. The bear grave was a wide-open pit with a few bones and small patches of fur remaining. We crawled into bed, after setting up a temporary camp, around eleven o'clock. TOTALLY EXHAUSTED!

Possible Camp #3

We cleared and extended the trail started last year to the claims. Bob planned to build another campsite but first we had to find a source of water. If we didn't find water in the area we would have to travel all the way back to Camp #2's source. We hiked, panned, and explored, all to no avail.

First Bear Encounter

I was preparing to start breakfast on the morning of our third day at Camp #2 when Bob stood up from where he was sitting and whispered, "Get the gun. A bear!"

I caught a glimpse of the small bear as it stood at the edge of the surrounding brush. I then realized as I rushed to the tent for the gun that I still had the breakfast foods in my arms and quickly set them down on the ground. I grabbed the rifle and hurried back to Bob. He was standing very still and staring at the bear that was approximately one hundred and fifty feet away. It, too, seemed to be transfixed in a gaze. As I handed Bob the rifle, without taking his eyes from the bear, he whispered, "Get the camera!" Again I rushed to the tent. By the time I returned to Bob's side, the bear had disappeared into the four to six foot high new growth brush which was just high enough to hide a bear!

I looked at him questionably and asked, "Why didn't you use the revolver? It was right next to you!"

He calmly answered, as he returned to his chair, "The rifle is the only weapon to use at that distance."

Now my knees shook and heart pounded as I thought to myself, you plan to cook bacon for breakfast!

But out loud I quickly said, "I am not going to cook the bacon."

He turned, looked at me and asked, "Why not?"

"I am hesitant," I explained. "The smell of bacon cooking will attract a bear back into camp."

"So what! Cook it!" he ordered.

I glanced over my shoulder and scanned the thick brush as I cooked breakfast. All the while I was thinking the bear would come out of nowhere for his breakfast or I should say our breakfast!

I was frustrated watching Bob nonchalantly sitting in camp with the portable radio in hand. As he searched for his favorite commentator, Paul Harvey, he looked as though he didn't have a care in the world. It angered me as I thought, "how could he be so calm after the bear episode!"

When I complained about his not seeming to care, he told me, "If it goes away and doesn't bother us, we really don't want to kill it."

We took a few precautionary measures before we left camp, in case the bear returned. I hung blankets on the clothesline in front of the tent, made sure Big Red was covered and all foods were stored in the barrels. To save gas, we both rode on the four wheeler. Everything

I did that day was with a glance over my shoulder and the surrounding area. As we traveled over the trail my thoughts turned to the recent bear attacks in British Columbia, Canada, where a woman had been killed by a black bear while hiking with her husband. There had been several encounters between humans and bears that summer in Alaska for the berries were not as abundant as the year before and the bears were hungry.

Four Wheeler Attacked

We ate lunch while sitting on a ridge on one of the mining claims when Bob decided it would be a good place for Camp #3. Leaving the four wheeler parked on the ridge, we headed down into a different valley in search of a water source for the new camp, one we could drive to.

We searched up and down the valley with no luck. We passed several raspberry patches with fresh bear sign as we began the steep climb back to the machine. After the incident of that morning, being more aware of the possibility of bears nearby, we made lots of noise and nonsensical talk. Bob acted nonchalant, although I noticed he was singing while banging the machete and shovel together quite often. When he sings, I know he too is nervous.

We checked several areas that would make suitable campsites as we neared the top of the ridge. I was ahead, deep in thought and oblivious to the surroundings, until I came to where the machine was parked. I realized something was wrong! There sat the machine with shredded foam scattered about on the ground.

Startled, I yelled to Bob, who was just coming around the ridge, "Bob! we've had a visitor!"

Both of us were flabbergasted and couldn't believe what we were looking at! OUR BRAND NEW MACHINE! With less than one hundred miles, it sat there with it's seat ripped apart, gas tank vent missing, teeth marks on the side, plus telltale black fur on the machine.

We picked the foam pieces up, placed them back onto the frame of the seat and headed back to camp, now aware of how thick the brush was in some sections along the trail. Each time we returned to camp after a day's excursion we wondered if there had been any visitors.

I already had doubts about making camp #3 in the area Bob had decided on as it was surrounded by trees and visibility would be limited because of thick underbrush. I tried to convince him that I was quite happy at Camp #2 with its open area. I reminded him of how many fresh bear signs were on the ridge where the claims are and the one before it. I was definitely sure I didn't want to make a Camp #3 after that day's episode.

More Samples

We spent several days in camp because of rain. I tested the soil samples we had gathered earlier and using the Geo-Chem process, found they formed a pattern on the map. Remember, the Geo-Chem process was explained at the beginning of the book. Each sample taken was marked on a map. Once a pattern had been established, the area would be rechecked by gathering and sampling more samples until a definite pattern was formed where the possibility of minerals, but not necessarily gold, could be found.

I dowsed while Bob panned on one of the sample gathering trips. We hadn't realized how far we had traveled down in the valley and knew we were in the lower part when the creek flattened out. There was an ice shelf several hundred feet long. Ice remains until late summer in the deep valley because of limited sunshine.

I was nervous as we followed a well-worn trail through the thick brush back along the creek and wondered if it was a fox, lynx, wolf, bear or moose trail. I surmised it was a bear trail as it was not high enough for a moose but not low enough for a fox, lynx or wolf.

When I told Bob what I thought, he looked at me, shook his head and snapped, "You think everything has something to do with bears!"

I kept alert as we continued to follow the trail and noticed Bob did too. We veered off the trail and climbed the steep hill up to the ridge where the machine was parked and were relieved to find it had been untouched.

"This time!" Bob commented

More Rain

The weather forecast was for more rain. We returned to

civilization after eight days of prospecting, as it was time to tend to the garden. After recent rains, the wild mushrooms were growing and needed to be gathered. Time for our salmon too!

I was ready for a good night's sleep after so many nights of waking with a start from the sound of a snort which was Bob snoring or the sound of a branch snapping. I had nightmares of bears in camp and coming after us. Time to go home!

The morning of July 24th, we left after packing all the supplies into two barrels. Bob loaded the barrel we had brought, back into the trailer! He laughed when I told him I thought he wanted me to have more practice towing a trailer with a barrel that would rock back and forth. But he said we had no need of another barrel at camp.

Bob spotted a porcupine in a small tree along the edge of the trail and asked if I needed quills for my beadwork. Yes! He knocked it to the ground and killed it with the shovel. It was not easy to pull the quills out but with a pair of long nose pliers we managed. It had the cutest face. Its feet resembled a baby's foot and the underside was coarse fur, not quills. It is assumed porcupines are slow so that if someone was hungry, and had no gun they could kill one with a stick.

The oil warning light on the four wheeler lit up just before the end of the trail, as it was quite warm that day with no breeze.

While waiting for the engine to cool, Bob said, "That is one thing I don't like about this machine. It needs a fan to cool the engine on a day like this and they are optional."

Third Trip

The morning of August 7th was clear, but the prediction was for more rain to begin later in the day. As long as we had camp set up before it began, we would be all right. We made good time as I was now used to driving the four wheeler plus the trailer had less weight and no barrel rocking back and forth. The four wheeler felt like a Cadillac compared to my Honda 185 that I drove for the last eleven years.

Another Bear Encounter

We stopped for lunch along Rion Lake. Just as we continued down the trail, I saw two black bear about two hundred yards ahead cross the trail. I quickly stopped, my pulse quickening as I told Bob who hadn't seen them.

"Ahh, they're gone by now," he said.

I informed him that he had promised we wouldn't take any chances as we were by ourselves. Then I sort of demanded, "Then we had better have our guns ready!"

He agreed, with that silly grin of his, like he was doing me a favor. Sometimes I think he was happy that I was the cautious and scared one. That way he wouldn't have to admit he was or he wouldn't want to scare me by admitting he was scared.

"Will you go first?" I asked.

"Nope!" he quickly answered. "You're leading the way this time."

I swallowed hard, squared my shoulders and pushed the throttle ever so slowly as I worked my way to where the bears had crossed. All the while my eyes searched the thick brush and spruce trees along the trail. I glanced over my shoulder several times to see where Bob was. I thought he seemed so far behind but I kept going, stopping only when I thought it was safe to wait for him.

As he approached me, he said, "I couldn't keep up with you!"

I laughed, as I returned the gun to its case and told him "Here I thought I had been traveling slow and cautiously."

With no trail to clear, we arrived at Camp #2 in just under five hours. No bears had been in camp.

After the bear sightings and fresh bear sign along the trail between camps and in the general area of camp, Bob suggested, "We should build a shelter around the tent area for a little more protection from critters." I was all for it!

We used fire killed trees to build a three sided "Fort" to the height of four and half feet. While Bob cut poles, I used the four wheeler to drag the larger poles for the base and smaller poles were carried by hand. Once we had a stockpile, Bob notched the ends, set them in place and secured them with spikes.

We set the tent up inside the shelter five hours later and parked the two machines in front of the open end. At that time there was darkness between midnight and 4:00 a.m. and being surrounded by walls gave me a sense of security without fearing a bear or moose would approach the tent in the dark of the night.

Quartz Rocks

Temperature reported on the radio the next morning in North Pole was 45 degrees. Although the sun was shining where we were, it felt much colder. We headed out early, anxious to get back to the claims to gather more samples. I was curious as to what was under one of the large quartz rock so instead of taking samples, I started to dig around it. That particular rock was approximately three feet by three feet. The surrounding ground consisted of large granite rocks wedged haphazardly together. I didn't get it completely dug out as it was not easy digging!

While I worked, Bob had left to cut a trail down towards the valley. He had the revolver and I, the rifle, which I set on the rock right in front of me. It was so quiet, even the rustle of a leaf sounded like an immense noise!

It was quite cool by the time we returned to camp. Bob went after water while I picked blueberries, which were at their peak. Yes, I had the rifle next to me, although I did get carried away and found myself at least thirty to forty feet away from where I had set it down. Mustn't do that!

A Day in Camp

Oh boy, did it rain all night! We had dug trenches around the shelter and placed extra tarps across the top to keep us warm and dry. With nothing to get up for, we slept until 10:30 a.m. when Bob decided he was hungry.

It was a cold, rainy day. The tarps covering the kitchen area created a draft, which drew the smoke our way. Bob built a teepee fire to radiate the heat to where we sat about seven feet from the fire pit. A teepee fire is built with the firewood set teepee style to create its own draft to draw the smoke up straight and away from us.

I did a little extra cooking. I made cornbread in the frying pan and cooked the wild mushrooms that Bob had picked. They looked like an edible kind and smelled good. They were an extra treat to the canned chicken, potatoes and vegetable dinner.

More Trail Making

Oh oh, we both had cramps and diarrhea, but felt fine otherwise. Now we knew which mushrooms were not edible! Near the "Dunny" we found the toilet paper pulled out from under the board we had protecting it from the rain. Pieces of it were scattered about. Most likely a squirrel or vole had done it.

We headed out to explore the steep ridge at the fork of another feeder creek where my dowsing rods had indicated something. According to the topographical map, the ridge we were on looked as though it could be traversed with the four wheeler. We cleared about eight tenths of a mile of trail then walked ahead to check what the terrain to the creek would be like. The area we came to was not where we wanted!

Following the ridge above the valley we recognized a burned area as the one above the section of the creek we wanted to explore. Bob marked the trail as we returned to the machine. We cleared trail for two and a half-hours to reach the burn area.

We heard a sound while eating our lunch as though something was traveling through the brush back up the trail. We listened intently, the adrenaline working, guns ready. Nothing more! "It may have been a branch or tree falling that we had cut as we made the trail," Bob remarked.

Each time we are scared by something like that one realizes you're not always alert to your surroundings and should be!

Camp Intruder

We slept quite well most of the night for we were tired from clearing trail. Nature called at 6:00 a.m. and as usual I checked the kitchen area. Everything was quiet, so I returned to bed and went back to sleep. Not for long! At 6:30 the sound of a branch snapping woke me up, but I thought, "Oh, I've heard that before and it never was anything." All of a sudden I heard the dishes rattle.

"Bob!" I whispered quietly, as I nudged him awake, "I heard the dishes rattle."

Groggily, he looked at me. I repeated that I had heard the dishes rattle.

We carefully unzipped the tent and from my side, as we lifted the flap to peek, I saw a small grizzly bear. It was rubbing its head in the dirt by the fire pit looking just as content as could be. First it turned its head one way, then the other, oblivious to our presence.

When we made noise as we dressed and readied the guns, he stopped, sat back on his hindquarters, ears perked. He looked down the trail. Apparently, he didn't know where the sound had come from.

The view I had of the bear looked as though it would be a good shot and told Bob to shoot, not realizing I had a better view of the bear.

"We must shoot it!" I nervously whispered. "Shoot! Shoot!" I shouted as the bear faced us.

It seemed like an eternity before he shot. Kaboom! I saw the bear falter and shake himself. With a look of disbelief on it's face it turned and meandered slowly away as we emerged from the tent. It stumbled several times.

Bob mumbled to himself, "It sounded like a killing shot and I hate to shoot it again. I don't want to ruin the meat."

He explained to me as we followed, waiting for it to go down, that brush had obstructed his sight for a good shot but took the chance. We turned to look at each other when the bear disappeared. In a matter of seconds, it was gone! We lost sight of him amongst the thick brush.

We realized we didn't have our shoes on and hurried back to finish dressing before following the trail of blood and warm droppings. On a clump of grass we found a drop of blood that contained berries and was already clotting. Was it coming from the throat or where?

The Revolver

We stayed close together, guns ready as a wounded bear is not something we wanted to tangle with unexpectedly. We lost the trail about two hundred feet below camp and continued to search for quite sometime but found nothing and returned to camp.

It was some time before my knees stopped shaking while Bob as usual, acted nonchalant which upset me.

Not having much experience with the .44 revolver I asked

Bob, "If I had shot the bear with the revolver would it have killed it?"

He thought for a minute then answered, "You really have to be close to a bear for it to be effective. In other words you carry one hoping not to have an encounter with a bear at close range. A rifle is good, a shotgun is good, but a .44 revolver is questionable."

"Oh great!" I answered. "Why did we get one?"

He snapped back, "So we wouldn't have to carry the rifle everywhere we hike. Most of the time the .44 would be effective."

Mama Bear and Her Cubs

With all the excitement of the early morning, we did get an earlier start to the claims. We constantly looked for some sign of the wounded bear as we traveled along the trail. I mentioned how dense the brush was along the burned over ridge.

Bob laughed and said, "You never had a wounded bear out there either."

"Yeah, you're right," I said. "Plus, we never had such a close encounter with a bear before."

We followed the new trail we cut the day before and continued it to the last steep ridge before the creek. We planned to work where two feeder creeks to Ahtel Creek joined. The ridge between them was an area that we had wanted to concentrate on. The year before, my dowsing rods indicated a strong pull for minerals.

In all the excitement of that morning I had forgotten the backpack. Leaving the lunch in the box of the four wheeler, we took the shovel, pail, gold pan and dowsing rods with us. Of course, the trusty .44 revolver too!

The ridge down to the creek from the four wheeler was very steep. Bob began to dig and pan gravel along the edge of the creek. I walked ahead thirty feet and set the pan down on a downed tree. I started to pull off my work gloves in order to work with the dowsing rods when something on the ridge across the creek caught my eye.

IT was a BLACK BEAR! It was on top of the fifty-foot ridge above the creek about one hundred and fifty feet away. I was speechless for a few moments and my heart began to pound.

I yelled, "Bob! A black bear!"

I think I screamed it.

A curious young cub peered over the edge of the ridge when I yelled. It quickly disappeared when suddenly, two young cubs were scurrying up a nearby tall spruce tree. What a racket they made moving so fast! In a matter of seconds they were at the top clinging to it.

The mama bear continued to stare down on us.

Bob banged the machete and shovel together while we loudly shouted, "Get out of here. We are not here to hurt you!"

When she vanished from the ridge, I realized I had unconsciously taken the .44 out of the holster and held it in my hand while my knees shook.

"She saw us. She should leave now," Bob quietly said.

I wanted to believe him, but wondered where she had disappeared. Would she continue to come after us, I wondered!

We Were Stalked

Suddenly, she appeared on the ridge directly across from us. Again we hollered and made noises but she slowly kept coming, all the while sniffing the air, trying to catch our scent. She would hesitate each time we yelled, but kept coming!

I looked down at the .44 in my hand and said, "G-D-,(I rarely ever swear) here we are with a small gun!" All the while the thoughts running through my mind were of the discussion that morning on how effective the revolver was in a situation such as this. I clearly remembered Bob's remark, "The revolver is only good at close range." I then had doubts as to the reliability of the .44. I was scared!

We watched her slowly move down the ridge stopping, sniffing the air, hesitating, then continuing towards us. I almost lost my balance, as I started to back away, when I bumped against a downed log hidden in the grass.

"Where is the gold pan?" Bob asked.

Hesitating, I nervously said, "It's on the downed log near the creek." I yelled, "Are you crazy? She is still approaching us!" as he went to retrieve it. He ignored me and kept his eyes on the bear.

As I backed up, I again almost fell backwards over another hidden log. I slowly retreated to the ridge we had come down. I noticed Bob was right beside me. We made more noise but all to no avail. SHE KEPT COMING!

Knowing the rules of not running from a bear or turning your back on one, I said, "I don't care if you're not suppose to turn your back on a bear, she won't know what side of me is facing her."
I turned quickly, made a short dash, stopped to check where she was and realized Bob, too, had done the same.

Together we quickly moved back another twenty feet or so, stopped and checked to see where she was. She, too, had stopped. That happened several times before we reached the base of the steep ridge and quickly began to climb it, with the bear still following. We stopped only long enough to glance back over our shoulder making sure she kept her distance.

At one time it appeared as though she was closing the gap between us. We usually stopped several times to catch our breath but didn't dare to this time. I thought my heart was going to come right out of my throat, as it was beating so fast!

Nervous, scared, exhausted and with the .44 in my hand I became concerned. I wondered if we had come up the ridge in the direction of the four wheeler where the rifle was. Again I swore (I really was scared!) and wondered under my breath, "Where's the G— D— four wheeler."

I thought it was under my breath. Then I heard Bob let out a chuckle. Still acting nonchalant, he said that he thought she was cute! I figured he did that to keep me calm and not let me know how scared he really was.

As I caught a glimpse of the four wheeler. I looked back down the ridge. I could see the bear was still climbing right behind us! We hurried to the machine. Bob quickly pulled the rifle from the case (Remember, it is the old British .303) and stood on the seat while I kept watch. All of a sudden I saw movement and it was BLACK!

I yelled. "She's coming over the ridge!"

Bob said, "Let's wait and see what she is going to do."

Then he quickly added, "She is still coming towards us so I had better shoot close to her to warn her."

KABOOM!

The sound rang out as he shot a short distance in front of her causing her to quickly retreat. We waited several minutes wondering if she would return. Nothing!

I was pretty shaken at that point and nervously said, "Let's get out of here! I don't care if that is the area of the lost Gold Mine. I want to get out of here! I want to go home!"

I hurriedly tried to pack things back into the box on the machine. They wouldn't go in right and I repeated, "I want to go home!"

Bob was laughing as he took over the packing for me. Usually I'm the packer, not him! As he packed I told him I would start up the trail ahead of him. I didn't want to ride up the steep section on the machine. He said, without looking at me and continued to pack the gear, "I wouldn't if I were you. We don't know if the bear went down the ridge or not. She could be up here circling us."

"Well! I'll walk behind you!" I snapped.

I kept up to him for the first one hundred feet or so as he had difficulty maneuvering the machine in the steep sections. When he realized that I had started to lag behind he waited for me. Once we got to where it was safe for me to ride on the machine, I gladly got on

The three-mile ride back to camp gave us a chance to discuss the situation that we were in. I insisted we return home today. He still wanted to go back to the same area to dig a few holes. I had him convinced by the time we got back to camp, that I was scared and ready to leave.

In my argument I had said, "There is no need to put our lives in danger just to look for gold. It is not worth it!"

While we prepared to close up camp Bob noted a few ravens flying just below the ridge of Camp #2.

He said, as he looked at me and shrugged his shoulders, "Two landed in a tree and the other disappeared. I bet the bear we shot this morning is down there. Dead or alive."

Then he admitted, "I'm chicken too. A wounded bear could be dangerous and our safety is much more important than salvaging the meat."

I too hated to leave, but two bear episodes in one day was a little too much for me. Bob understood. Maybe he was scared also?

I led the way back and when we arrived at the Luv four hours and fifteen minutes later, a record time for us, Bob said he had a hard time keeping up with me. I didn't tell him I was anxious to get back to safety. Instead, I mentioned how driving the four wheeler felt like driving a Cadillac! With an empty trailer, traveling was easier.

We were thankful to be safely back, but disappointed in not finding the gold. We planned to return in late August. We were busy preparing for hunting and winter so we didn't return as planned. I was still paranoid about bears and was happy we couldn't return. We kept thinking the gold had to be somewhere nearby Camp #3 and planned to search again in summer of 1997.

CHAPTER FOUR

Our Third Year

There is an old saying, "Third Time Never Fails." We hoped it was true. We would know by the end of summer of 1997.

We survived winter of 96/97 in Alaska. And enjoyed every bit of it, even experiencing minus fifty-two degrees! That was our second winter in Alaska. The first in 84/85 was spent in an old log cabin in Sourdough, Alaska located at mile 142 ¾ Richardson Highway. It was rugged and a real learning experience. Wilderness living and surviving minus fifty below was our challenge. Oh, so many experiences, I would love to share with you.

With the sun barely rising above the horizon throughout the mid winter months, its long shadows and low rays created many hues of pink, red, violet and blue. These mixed with the white snow against the Alaska Range in the background, capped with azure blue sky above. It was awesome! Sunsets were gorgeous!

Snow machining, ice fishing, visiting and just being lazy was as Bob would say "NTTTT" which means, Not Too Tough To Take. Dick Albert, the weatherman on Channel five in Boston, Mass. used the saying when describing nice weather.

In May, we attended a Rock and Mineral Identification course offered at the University of Alaska Fairbanks. We needed to understand the many rocks and formations we found the summer before on our claims.

After I had experienced how easy driving a four wheeler over the rough trail to the claims had been in comparison to the three wheeler, I insisted we purchase a second one for Bob. Again, moneywise it was not feasible but safety-wise it was a necessity. The real reason may have been that I didn't want to pull the trailer again because I was the only one with a four wheeler.

Plans were to spend most of June working on the claims, but first we had to plant the garden and finish some main projects, like staining the house, gathering food and supplies we needed to take out with us and also preparing for company in July. We also made time to travel to Fairbanks three nights a week to attend the Rock and Mineral Course.

The First Trip

It's nice to be loved! Our neighbors, Skip and Lin Peters, were concerned for our safety out in an area where there were lots of bears. They offered us the use of their cellular phone, although they weren't sure if it would reach from so far out. Cellular phone usage is limited in Alaska because of the wide range of wilderness.

Bob didn't like the idea but I, on the other hand, felt that I wanted some security after so many bear episodes last summer. To convince him, I told him they were curious as to whether or not it would work from that distance and they wanted us to test it.

The Trail

Some sections of the trail had more mud than we expected, although there had been a drought. It was caused from the thawing of the frost in the ground. We cut several detours in areas that we had driven through last year.

Bob towed the trailer with the 96 Suzuki while I had the honor of driving the brand new 97 Suzuki! He tried to use the super low speed and differential lock, which make all four wheels work independently, when he became stuck in one mud hole. It still would not move forward or backwards. He attached a tow cable to my machine and with both working together, we managed to pull it out. Instead of a winch on the machines, we use a cable for such emergencies.

When I mentioned, I had noticed fresh bear scat and several very large prints along the trail around the lake, his usual response was, "It was a small black bear." Deep down, I knew better! The trail on the first trip of the year needs to be cleared of winter blow down. Between cutting detours and clearing brush it took two hours to clear the four-mile trail around the lake. Once, on the main trail, it was travel as usual with only one detour to cut.

A large cow moose was feeding at the further end of Rion Lake. I noticed, as we passed the old trapper's cabin near the lake how it had deteriorated tremendously in the last three years.

The bridge, we helped Tim and Eva build last July, was a little shaky. I asked Bob if he had noticed.

"I noticed one of the poles bracing the bridge had its banking undercut from the high water in spring," he explained that it should be solid enough to withstand travel over it.

The PRIVATE PROPERTY NO TRESPASSING sign was still in the middle of the trail that led to the trapper's cabin. We veered off onto the by-pass trail to join the main trail above his property.

After a few minutes on the main trail, I smelled smoke like burning grass and thought it was coming from the trapper's wood fire. It continued, so I yelled to Bob to stop and told him about it. He, too, had smelled it. It became fairly strong as we stood there, so decided to check the machines. It was coming from Bob's. When traveling through muddy areas the leaves and grasses get stuck behind the screen that protects the exhaust system. They were smoldering! We waited for the smoke to subside. As we continued down the trail we smelled the smoke again. Bob turned the fan on that he had installed to cool the engine. It stopped smoking. Remember, last year we had to stop several times with the new 96 Suzuki to let the engine cool because it had no fan. Well, he installed one on that machine.

The trail was fairly clear with a few large downed trees across it, which we detoured around instead of cutting through. Squaw Peak was not a problem anymore now that we both had a four wheeler. However, the trail down the backside was still rough. Bob recut some of the stumps in the trail. Each time we traveled over the trail the moss was pressed down exposing more of the stump. As Bob took the chain saw out of the box in front of his machine, he said, "Got to make a good trail!"

After last year's bear episodes I realized I had become paranoid about meeting one on the trail. Several times, through certain sections of the trail, the tires would make a different sound and it would startle me. Even a dark spot in the woods ahead or on the trail would resemble an animal. This got my adrenaline going, although I knew that it most likely was a tree or downed root system.

Camp #1

We arrived at Camp #1 after six long hours on the trail and found that it had been visited by bears. The yellow barrel was lying on its side about eight feet from the tree to which it had been chained. Although the bear or bears had broken the cable they had not broken through the tough plastic of the barrel. They tried! There were plenty of gouges carved into its side from their claws. The remaining metal barrel with the locking cover was still upright and chained to the tree although the ground beneath it had been dug out. That must have been one very frustrated bear or bears.

Strangers at Camp #2

After eight hours on the trail, my shoulders were sore. My hands felt as though they were molded to the grips on the handlebars. I then realized, for the last few hours, I had become oblivious to my surroundings. Lost in my own thoughts I had automatically responded to the rough ride and focused solely on the trail.

I wondered what we would find as we approached Camp #2. No bears had visited. As we checked the area, Bob noticed fresh footprints in the dust surrounding the fire pit and two empty MRE packets laid in it. The two metal folding chairs were down from where we had hung them last fall. No one had slept in the shelter area but someone had checked it out. Who were they?

I mentioned to Bob, as I prepared lunch, that Tim and Eva returned to pick berries and hunt last fall and it could have been them. But Bob, deep in thought said, "No, these are fresh since the rain of a few days ago." Then he added, "Who were they?"

I tested the cellular phone from Camp #2 which is in line with a transmitter tower located on Donnelly Dome and left a short message on Lin's answering machine. Now I had a sense of security knowing I could reach someone in case of an emergency.

We kept a smoky fire burning throughout the evening to keep the large furry spring mosquitoes at bay. The wind began to blow during the early morning hours. Each gust of wind would whip the tent like a balloon. It billowed out, snapped, then, deflated. It kept us awake.

We cleared lots of fallen trees between camp and the mining claims. We checked the helicopter pad where the trail crosses to see if whoever had been in camp had come by helicopter. We couldn't tell, but did find a few crackers and a brownie from an MRE packet that had been left on a stump. They had not been rained on nor attracted any animals.

I heard an eerie sound as the wind howled through the fire-killed trees while we rested, and thought to myself, "It must be the old prospector welcoming us back on our third attempt to find the lost gold mine."

A few moments later Bob said, "Maybe the old prospector is coming to tell us where the gold is."

I laughed and told him what my thoughts had been.

"After forty years of marriage it shows we think alike!" he remarked.

We finished clearing trail, gathered a few soil and rock samples, then returned to camp.

The Skull

Later that afternoon Bob asked if I would go with him to search for any remains of the grizzly bear that he had wounded in Camp on August 11, 1996. We searched the same spot we had seen the ravens circling the ridge below camp as we were preparing to leave after two bear encounters.

We walked fifty feet apart and on the third pass Bob found the remains of a skull and surmised it was from the bear he had shot. Nearby, in an area of five square feet, we found one knucklebone and a few scattered clumps of fur but no claws. Everything else was gone! Bob was happy to know he had killed his first bear.

A Night Critter

We found the water source of Camp #2 barely a trickle. It tasted all right although it was very yellowish in color. My theory about "less water," was that the ground under the moss had not thawed yet and by July it would be running like last year. But, oh no, Bob said it was caused by the dry weather and that the stream comes from runoff. We'll see!

Using the small can left by the stream last year, Bob patiently scooped four gallons of water into a container. Again, I walked up the steep hill.

It was quite warm that night. We left the tent flaps open and parked the machines in front of the shelter.

I awoke with a start around 12:30 a.m. when Bob sat up quickly! Before he said anything I heard the crunching sound and whispered, "What is it?"

"I think its a porcupine chewing on the new machine," he answered.

The sound stopped. I put my hand on the revolver by my head while Bob got the camcorder ready. Through the screen, I saw a funny little face peeking around the front tire of the new machine. It gave me the impression, as it quizzically looked at me that it was wondering "Who are you? What are you doing here?"

The porcupine looked up as Bob emerged from the tent then backed away a foot or two, turned and looked back in defiance. Bob asked, "Do you want the meat?" but quickly added, "You will have to skin it!"

"No thanks!" I responded quickly without hesitating.
If he had volunteered to skin it, I would have loved to have the meat. We have eaten porcupine and found it similar to pork.

It had a look of disgust on its face as it slowly meandered away as if in slow motion. It stopped several times turned to see what we were doing. Each time it stopped, I clapped my hands and shouted, "Get moving!"

Bob picked up a small piece of wood to throw at it. He hit it in the butt. The porcupine slowly turned, looked at him and began to chatter noisily before disappearing into the bushes, still at a slow pace. It is their nature to move slowly, as they have no enemies but man. Bob showed me the wood he had thrown which now had several quills embedded in it. We couldn't see where it had chewed on the machine because of the diminishing light at 12:30 a.m. The rest of the night we slept restlessly as the wind began to blow again. By 6:30 a.m., I could not stand the tent whipping with each gust of wind. I told Bob, "I'm going to get up, but don't want to start a fire until you get up."

"I don't think we should start one as it's too dry and the wind is blowing hard. Let's pack up and leave early." Surprised at what he said! I responded with, "You mean, we have to do without our morning coffee! What are we going to have for breakfast?"

"Peanut butter, crax and tang," he suggested. Yukkkkk! I

thought, but had to agree it was best not to start a fire in those conditions. Traveling on the four wheelers was so much easier and quicker and we made it back to the Luv in four and a half-hours.

Second Trip

After five days at home preparing for company, on June 10th we headed to the claims although there was a threat of rain. In some areas of the trail the mud was thick and soupy. We detoured through places we had driven through the week before.

When Bob got the machine and trailer stuck I worked my machine through the woods around the mud hole to pull him out. The rest of the trail, other than that, was not bad and cut only two detours. These were now detours around detours that had become wet and muddy themselves.

The trailer contained another barrel, which did not rock back and forth. Bob made sure of that! It also carried a stove, several 2 x 6's for door jambs and window frames, boards and slats to build a ladder, plus five, one gallon cans of gas. Progress was slow. We stopped several times to secure the load. With no rest stops or stopping at Camp #1, we arrived early evening at Camp #2. We guesstimated the time for I had forgotten my watch.

With no bear sign along the way or at camp, we slept quite

Figure 6: Hauling building supplies for camp #3.

well. I awoke several times during the night and with twenty-four hours of light, I kept thinking it was time to get up. I decided to wait until Bob was ready.

He used the Global Positioning System (GPS) to check the time while I prepared breakfast.

A few minutes later he chuckled and said, "Guess we needed the long rest. It's 10:42 a.m."

Camp #3's Beginning

We cleared brush from the area where we planned to build the log cabin. While I piled brush, Bob cut several large dead trees for the base logs and used the four wheeler to skid them. He scribed and set the logs in place to form the base. He skidded several more logs before we left for the next morning.

The next morning we were hard at work by 10:00 a.m. I wanted to help with the logs but he felt it was safer for him to skid and handle the large logs. I set up a comfortable campsite.

The stove Bob had built was set up in the cleared area. I took great care in preparing a fire pit. To form a good base you remove the topsoil and roots to a gravel or sand base. If the roots caught fire, they would slowly smolder underground for many hours igniting into a possible forest fire.

"Ruth Nash would be proud of me!" I shouted to Bob.
Ruth was a Girl Scout leader trainer during my years as a Girl Scout leader. She was very strict as to how and where a fire pit should be built. "Thanks Ruth!"

My Jobs

While digging the outhouse hole I set aside a few quartz rocks for testing. When I finished, Bob told me to assemble "the throne" seat.

"Me? You do the carpenter work!" I exclaimed.

"It's about time you learned!" he snapped back.

I took the four pieces he had precut for the dunny and knew the one with the hole in the center went on top. I arranged the other three pieces beneath it. I asked him if it were right before I nailed it.

"It's so simple, anyone could put it together," he said as he approached.

It was wrong! I had two pieces backward. Hey! Two out of four ain't bad! Then he told me to use small nails that were in a can in the barrel. Well, I checked every can and found the smallest were eight-penny. I thought they were rather large but started to use them anyway. After all, he told me they were in the barrel.

As I began to nail it he yelled, "Don't forget to have a backing when you nail two boards together!"

"I know that!" I snapped, exasperated that he thought I was so dumb!

After much frustration, I found smaller nails. I assembled it then set it over the hole I had dug. The hole was deeper than others we dug therefore, the hole was larger around. Oh no, it barely covered the hole! I sat on it, adjusted it this way and that way until it felt secure. I realized as I looked around that there was quite a view from the dunny.

We kept a small smoky fire burning in the fire pit most of the day to keep the mosquitoes at bay. I lit a fire in the stove to cook lunch and decided to have MRE's so there would be no cleanup as the main packet is heated in boiling water. They may be convenient but with the extra weight of the individually packaged packets for a single meal, a person could carry several meals in comparison to weight for four people. Bob complained that afternoon that he kept burping up the taste of plastic and was thirsty. We decided that would be our last MRE meal.

Work progressed throughout the day. I began to chink between the logs with moss. At first I gathered it close to the cabin. I used the pulaski, which is a light weight grub hoe used by fire fighting crews, to cut the moss away from the roots and dirt. I packed the moss into two-grain bags and carried, maybe dragged them, two hundred feet or so to the cabin. Bob had done a good scribing job in fitting the logs tightly together. To pack the moss in the spaces, I used a large serving spoon. My tool! The reason for gathering moss close to the cabin was that I was afraid to go further away because of bears.

I noted Bob upset over something and understood why when he told me, "The new Husquavana 51 chainsaw has no power and isn't working properly. Although I know what the problem is I don't want to work on it as it is still under warranty. I should have brought the other chainsaw but I figured that with a new one I would have no problems."

"Are you going to pull a Tim?" I asked. He scowled at me, so I sarcastically remarked, "Guess not!" The reason for my comment was that Tim had been known to throw a chainsaw he was having problems with. Not Bob, to him things cost too much to destroy in anger.

Small Critters

We watched a robin hop on the ground near the brush where bugs congregated. (Remember there are no worms in Interior Alaska.) With a mouthful of bugs, it flew to a branch in a nearby dead tree. A moment or two later it plunged towards the ground. Curious as to where her nest might be, we searched through the brush near the tree but couldn't find it. Apparently, birds like the protection of the new growth in a burn area. We had observed several species among the brush recently, including a grouse with young ones.

Later, as I read by the campfire, Bob shouted, "There's a bear in the tree!"My heart skipped a beat as I anxiously looked in the direction he was and saw a porcupine climbing the same dead tree the robin had been in.

Bob thought it was funny how I reacted and said, "You believed me!"

"Yes, you know I am still paranoid about bears since our episodes of last summer and I don't think it was funny!" I snapped back.

We walked over to the tree to get a closer look. The poor thing didn't know which way to go. It climbed higher and balanced itself on the dead branches and peered down on us. It would start to descend. Bob would tease it with a stick and it climbed again. Finally it slowly descended the tree using its tail for balance and to feel for a branch below him. It immediately climbed down the tree trunk. Once we left it disappeared into the brush.

Thunderheads were forming in the late evening, western sky and luckily they missed us. We had little protection from a storm with camp set on an open ridge. It became quite cool once the storm passed through and we retreated to the tent. Bob parked both machines in front, in case the porcupine should return.

"Let's hope he stays away from camp tonight," he said.

A So Called Day of Rest

We needed a day of rest! Sooooooo, headed to Camp #3 planning to collect more soil and rock samples. It was quite impressive as we approached camp to see our accomplishment with half the cabin walls set in place and chinked.

We cleared and flagged the boundaries of the claims using the GPS, (Global Positioning System) to locate the longitude/latitude. We posted claim papers at each Northeast corner of the six claims and returned to camp three hours later and ate the lunch I had carried all morning.

We hiked towards the dome, approximately three quarters of a mile to the southeast of camp in the early afternoon, to gather soil and rock samples. We had tested samples from that area but with more knowledge after attending the Rock and Mineral course, we wanted to recheck it.

Another Visitor

So much for our day of rest! Back at Camp #2 as I began to prepare supper, I noticed a ziplok bag on the ground, which appeared to have been shredded by something. I thought nothing of it until Bob called me over to the old garbage pit where he pointed out tracks that resembled bear tracks.

I closely observed them and said, "I think they are tire tracks" and returned to the kitchen area. I noticed the can I had washed and set on the shelf was now on the ground. The plastic jar containing distilled water, used for geo-chem processing, had been chewed.

"We did have a visitor today!" I excitedly said.

"Leave the dirty dishes to attract the bear into camp and maybe we'll get us a bear tonight!" Bob suggested,

Later, while he showered, he said, "There's your bear! Get the camera!" I caught a glimpse of a small brown bear just as it disappeared into the brush. I quickly took the rifle out of the case on the four wheeler and holding it tightly, I watched for movement in the surrounding brush. With Bob barefoot and naked, someone had to be prepared.

"You should have grabbed the camera, not the gun!" he said

angrily as he hurriedly dressed.

"But the bear disappeared before I had the gun out." I explained while I continued to watch the brush.

"Did you cock it?" he asked.

In my haste to be prepared in case of a confrontation with the bear, I had forgotten. Then he reminded me, "If it isn't cocked it would just go "click" when you pulled the trigger." Then he asked, "What would you have done if the bear came charging into camp?"

"Panic!" I sheepishly answered.

In the early morning hours of June 14th, a strong gusty wind again began to blow. We heard it coming through the brush before it whipped the tarps over the tent and kitchen area. A Whoomp! A Whoomp! it would snap, then settle down for four or five minutes only to be repeated with the next gust of wind.

I was up by 6:00 a.m. I couldn't stand the noise anymore. Bob followed shortly after and as he crawled out of the tent, he said, "I expected the bear to return during the night so I didn't sleep well. Plus the wind kept me awake." I had no sympathy for him. That was probably the first time he actually listened for bears! He had suddenly sat up during the night when he thought he heard something. It startled me.

We closed up camp. Ate a cold breakfast and by 7:30 a.m. we headed home. Bob, who never returns with an empty trailer, cut burls along the way. The trail was still muddy, but we had no problems. The detours we cut paid off. We arrived five and a half-hours later at the vehicle.

Company

We had lots to do before returning to the claims. I washed clothes the day we returned. It is not an easy job when you don't have running water or electricity. Bob gathered the supplies he needed to bring out and I repacked other supplies.

Our first company of the summer arrived. Ted Davis and Tom Merwin drove forty three hundred miles in eight days from Lakeville, Connecticut to Alaska. We left mining equipment for them to pick up as we did not expect to be home when they arrived. Bob and Tom went to Delta to have the chainsaw fixed. It was exactly what he thought the problem was. When leaving for the claims two days later,

Lin offered her cellular phone and said she felt better knowing we could call for help. I agreed.

We were on the trail by noon. The wild roses were in bloom and many different varieties of wild flowers along the trail were ready to burst forth with their beauty. I had found a miniature moccasin flower on an earlier trip. It is better known as a Lady Slipper. Its leaves and flower were a third the size of those found in New England. Oh, it smelled so heavenly!

We loaded a barrel from Camp #1 into the trailer. Upon our arrival at Camp #2, we discovered it had been visited by bears again! Bob had his mind set on moving to Camp #3. I was hesitant. I felt much safer at our established camp, which had been home for two years. It was comfortable and fewer trees surrounded it. I wanted to finish the cabin before moving but had no choice. I followed him to the new camp.

I noted when I set the tent up inside the cabin walls that they were not as high as the three-sided shelter at the other camp. We didn't set a tarp over the top because of clear weather. I sure didn't sleep well with only three-foot high walls surrounding us. I had nightmares about bears and was awake most of the night. The wind began to blow in the early morning hours but didn't affect the tent because of Camp #3's location on a wooded ridge-facing northwest, which protected it.

The sun shone on the tent at 3:45 a.m. but we slept until seven. We were ready to start the day after a hearty breakfast. I removed the tent so Bob could work on the walls. I continued to set up camp while he used the four wheeler to cut and skid logs. The logs he cut were from fire killed trees which still stood after a forest fire.

I started to chink the walls by late afternoon when I realized that I had to go further from camp to gather the moss. I casually mentioned that I didn't want to go by myself.

Bob turned away from what he was doing, looked at me with that scowl of his and snapped, "You're too paranoid! How many times have you slept in a tent? You never were this paranoid!"

"After being stalked by the black bear which really scared me, how can you blame me?" I defiantly responded.

He sounded disgusted with me when he asked, "What do you think we have guns for?"

I thought about what he said and as he turned back to work, I knew I had to do my share in building the cabin.

With determination, I strapped the .44 on, hooked the trailer to the machine, grabbed two-grain bags and the pulaski and boldly headed out! I caught a glimpse of him watching me go but as soon as I was out of his sight, I became nervous and aware of everything surrounding me.

I stopped at the first patch of moss I came to, approximately a thousand feet from camp. Not wanting to go any further from camp, I unhitched the trailer and turned the four wheeler around in the narrow trail.

I found a better patch of moss on the way back. I dumped the moss from the grain bags into the trailer and refilled them. On the way back to camp I thought to myself, "I did it! Although it wasn't very far I went off by myself!"

We traveled to Camp #2 for more supplies and water which still was a small trickle seeping out of the moss. Bob filled the containers, including the shower can, from a small water hole, which quickly emptied, as the trickle was not fast enough to replenish it. Maybe the water is from runoff and not the thawing ground beneath the moss as I had thought. Oh, well! We forgot one container cover so we duct taped a piece of plastic over the top. Most of the water was lost on the long trip back over the trail. It was time to search for a water source for Camp #3!

I set the tent up inside the walls, which now were five feet high. Bob parked the two machines in front of the open doorway for extra protection if a bear should come into camp. We slept quite well after an exhausting day, plus felt safer with the extra security of the higher walls.

The Roof

Bob cut and hauled logs the next morning. He framed the windows with boards as he set the logs. While he was preoccupied with his work, I did morning chores then chinked walls the rest of the day. Bob cut roof poles two and three inches in diameter as he cleared an opening for a view. I carried them to the cabin and aligned them by alternating one small end then one large end so they would remain straight as they were set across the roof. The evening became quite cool and it began to sprinkle. We set a tarp over the four walls and crawled into a nice warm, dry tent.

It was June twenty first, the longest day of the year with twenty-one hours and forty-nine minutes of daylight between sunrise and sunset.

In the early morning hours, I watched a spider crawl slowly up the back inside wall of the tent. Several small black flies and a few mosquitoes congregated near the peak. He positioned himself into the very peak and remained still. The bugs seemed to be oblivious to its presence. He missed several times trying to capture one. Once I saw him capture his meal, I went back to sleep. My morning entertainment!

Figure 7: The cabin.

Enjoying our morning coffee and sitting in the clearing with a view over looking Ahtel and Koda Creeks, we could see the trail along Ahtel creek and hoped someday to explore it and that valley.

It was time to start working. Bob trimmed the tips of the small poles at an angle to fit tightly against the ridge pole. I slid each pole up to him. He set each one in place on the roof in as straight a line as possible before nailing them. Over these we used moss gathered the day before to form a cushion for the roof tarp. It was time consuming and by ten p.m., the base of the roof was completed.

We returned home the next day. More rain was predicted. We had lots of home chores to do.

Fourth Trip

Lin hugged and kissed me as she handed me her phone and said, "We will pray for your safety."

We met up with a road crew replacing a culvert within the first mile of leaving the vehicle. We waited for their lunch break to cross the open ditch. I saw them watching us as we worked our machines through the deep sand and up the further banking. I had to admit the fully loaded trailer with lumber hanging out the back and a foam mattress tied on top of the load appeared strange, probably humorous.

We observed the bull moose in his usual area of Rion Lake as we ate lunch. We had seen him each time we traveled past the lake during the summer.

A cow moose stood in the water at the far side of the lake. Two white swans swam gracefully near the center. A muskrat swam near the beaver house while a female duck, followed by twelve little ones, swam away from the nearby shore. The trail along the lake and marshy areas was graced with the color of purple from the blossoming wild irises.

I ran over a stick just before the new bridge. It poked through the hard rubber along the edge of the fender. I stopped, yelled to Bob but he didn't hear me. Maybe he pretended not to. I struggled to pull it out. When I caught up to him he was video-taping a grouse in a tree.

"Didn't you wonder where I was?" I asked.

He looked at me and quietly said, "No, I figured on the four wheeler you were all right."

I told him what happened. He shrugged his shoulders.

"You managed didn't you?" he remarked.

As we traversed the turn onto the by-pass trail, one tire of the trailer ran up a two-inch spruce sapling, bending it down as it rode over it. I thought sure the trailer was going to go over when it tipped up to a forty five-degree angle. It righted itself after it passed over the tree. Luckily, there was no damage to the four wheeler or trailer.

The red warning light on the new four wheeler lit up as we descended Squaw Peak, which indicated a hot engine. We continued on a few minutes later when it had cool. My hands felt permanently molded to the handlebars when we arrived at the cabin around seven thirty. I felt like a zombie. I was so tired.

Another Visitor

Oh oh, a bear had been in camp again. We found the grain bag left laying on the ground in the cabin, out in the kitchen area. The board, with nails protruding up, set in the doorway was tipped on its side. Apparently the bear had stepped on it, but missed the rake Bob placed on the floor with its tines pointing upward. If a bear stepped on it, the handle would rise up, hit it in the face and scare it away. One window screen was torn and the large tarp over the kitchen area had been disturbed. And of all things, the outhouse seat was tipped over and the hole dug out. Yuk!

We worked the whole evening. Bob built a bed frame for the foam mattress, which was quite comfortable. When ice fishing at Birch Lake in the late spring we removed the foam pad off the melting ice. We couldn't pass up a bargain!

Bob screened the windows and doorway, keeping most of the mosquitoes out. We set a pic coil in a holder on the floor, oops, ground under the bed for the smoke to rise and fill the cabin.

I felt secure with the two machines parked in front of the open doorway. I was awakened at 3:15 a.m. by what I thought was the sound of an animal. But heard nothing more. I noticed a red glow on the cabin wall from the rising sun. The eastern sky was a glorious sight, aglow with crimson red and the sun barely peeking over the mountains of the Tanana Hills.

The next morning Bob attached wire from shopping cart sections on the outside of the windows. He had salvaged the shopping carts from a dump several years ago. One section of our yard is full of his salvaged materials which I called junk! I had to admit I was impressed when he used a lot of the materials on the cabin.

Building a Ridge Trail

We left camp at two o'clock and followed our trail of last year to continue cutting the trail to the creek. I was nervous, recalling the encounter with the black bear sow and her cubs in the area last fall.

We laboriously worked in the hot afternoon sun to level the trail, all the while, I looked over my shoulder and around the area as I expected the sow to appear.

Bob used the trench shovel like a hoe and I the pulaski to level the trail. We cut and pulled clumps of grass and dirt from the upper part of the trail to the outer edge, completing approximately seventy five feet of trail that day.

When I turned my machine around on the narrow trail at the top of the newly constructed trail, I watched Bob drive down it and end up against a tree because of the soft dirt. He tried to maneuver the machine back onto the trail but was making the predicament worse. When he noticed I had watched him, he motioned me to back down the trail. I hesitated. How dumb! Going down the trail before it had settled. I knew he needed to be pulled up. I reluctantly backed down but only as far as the long cable reached. It was amazing how easily it pulled with both machines in Super Low and Differential Lock.

"When it is in differential lock, the tires all turn at the same speed and work together. In four wheel drive, the tires work independently," Bob explained after I had asked him about it.

Back at camp around 5:30 p.m., we again had hamburgers for supper. NTTTT! (Not to tough to take). Bob worked inside the cabin most of the evening, putting up shelves while I gathered moss. I was a little braver, but still scared to be alone away from camp.

Water

The next day we worked hard and steady for several hours to finish the last steep section of trail to the bottom of the steep ridge. We found the main feeder running clear and cold. It tasted good. To scoop water, Bob dug a hole in the gravel of the creek. A water source!

Bob set poles along the edge and top of the roof to hold the clumps of sod in place. I packed moss atop the roof.

The next day was gorgeous! A cool fifty-eight degrees in the morning to eighty by late afternoon. Bob cut squares of sod thirty feet from the cabin then carried them up the ladder to place on the roof. Not being able to help him, I dug a hole in one corner of the cabin to set a five-gallon bucket in to keep food cold. Our refrigerator!

Satisfied that we had accomplished a lot in the short time at the claims, we returned home. The garden and yard needed to be tended before company arrived. Most visitors were expected around the fourth of July.

One Month Later

We left our company camped in our yard and on the morning of July twenty seventh, we left for our fifth trip to the claims. It felt good to be on the trail, which was now ablaze with the bright pink of the fireweed blossoms. The fireweed flower gradually blooms from the lower part of a twelve-inch stem. In mid-June, the two buds at the base of the stem bloom and each bud along the stem blooms gradually. The remaining two buds bloom by mid-August which indicates the arrival of fall. How true it is.

We gathered the abundant Boletis mushrooms along the trail for an extra treat with our meals. We know these are edible! The bear droppings near Camp #1 appeared to be a few days old but were fresh near Camp #3. They really checked it out! We found the black reinforced plastic on the roof clawed. It appeared the bear had climbed onto the roof. The squirrels or voles had pulled out some chinking between the lower logs. Other than that, everything was fine.

The Intruders

The next morning Bob set a water trough at the new water source. He dug a hole in the creek bottom then set the trough made from two boards nailed together in a "V" shape. The water flowed through the trough and could fill a container without stirring up the sediment from the creek bottom.

We headed downstream to check the intruders camp. One of the men involved in the camp had called and said they, too, were searching for the lost gold mine. They brought supplies in during the winter on snow machines and in June had landed his plane on one of the lakes in the flats. Hiking to their camp along the old bulldozed trail along Ahtel Creek was difficult. He said they had found no sign of a mine or any gold in the creek. They did find what appeared to be an old drift mine at the mouth of the creek. They would drop their prospecting claims in the fall if nothing was found.

A red and white striped box from their cache, high in the trees was on the ground. One of the three cache trees had claw marks engraved in it where the bear climbed to reach the cache. Luckily, it pulled down the box with the least amount of supplies.

We continued to follow the creek for some serious prospecting. Bob dug wherever my dowsing rods crossed, for samples plus enough to pan. Most of the holes were frozen mud, or no frost or bedrock. We came upon newly placed flagging tape tied to bushes and trees along the creek. The notches on several trees continued from where we had placed our markers. We filed more prospecting claims in the spring and wondered if we had filed over their claims or vise versa.

We filled the water containers at the trough and slowly crept up the steep trail with the machines in super low range to return to camp.

We set out in the late afternoon, in search of more large quartz rocks. There weren't any on the east side of the ridge but found several on the west side. We removed the material near one of the larger rocks for test samples. We thought the potential for gold in the area, was very promising.

Another Scare

I heard a branch snap around 5:30 a.m. on the morning of July twenty ninth, and thought nothing of it as I'd heard that sound before. I fell asleep but not for long! Suddenly, I was awakened by a scratching sound! Close by! I cautiously looked out the window. I had come face to face with the bear! It stood on its hind legs and was scratching the black plastic lining on the roof. At that moment, I was thankful for solid walls and bars over the windows!

Bob didn't realize a bear was nearby when I had awakened him. Without moving or taking my eyes off the bear, I whispered, "It's a Black bear!" He jumped out of bed, grabbed the rifle and headed for the door!

"Don't! " I warned, "The bear is right outside the door!"

"I am only going to open the peephole," he whispered. The night before he had cut a peephole in the door to put the gun barrel through. If a bear should wander into camp while we were inside, we could still defend ourselves.

I had to strain to look out the high small window located on the front wall by the door. I kept Bob informed as to the whereabouts of the bear. His view of the yard through the peephole was limited.

The bear moved to the side of the cabin. It seemed to sense something different. Maybe it heard us whispering. I cautiously peeked out the window. It was looking towards the front of the cabin. It hesitated, then moved slowly towards the four wheelers. It stopped. and listened, with its head cocked to the side.

It began to walk back and forth but stayed close to the machines. I wondered why Bob wasn't shooting. I watched him as he stood there naked and barefoot shifting back and forth trying to catch sight of the bear through the peephole. I didn't know at the time that the four wheelers were in his line of fire. He shot when the bear slowly headed down the trail.

I heard the thud of the bullet striking the bear. It stumbled, grunted angrily then dashed down the trail. It turned abruptly and came rushing towards the cabin before it lurched into the brush where it began to thrash about on the ground. A moment or two later, there was silence!

"You got it!" I yelled!

The adrenaline raced through our bodies as we quickly dressed. I began to giggle as I recalled the sight of Bob as he stood stark naked with the gun barrel poking through the peephole as he waited for the chance to shoot. What a sight!

The Bloody Trail

We cautiously moved through the brush to search for the bear. Blood covered the ground where it had thrashed but no bear! We looked at each in disbelief. Bob sort of mumble to himself, "I know it was a killing shot from the sound when the bullet hit."

He turned to me and said. "Let's go look for it."

My knees shook. Although I carried the .44 revolver I stayed close to Bob. I mean close! Bob pointed out drops of blood on a leaf or blade of grass as we followed the trail through the thick brush. The bear had left a fresh trail of blood each time it stopped. It was easy to follow the trail of compressed grass.

We lost the trail after following it for a half mile. We circled back through the thick brush. All to no avail. We returned to camp with the intention of searching later that morning.

There were fresh moose tracks in the yard. Bob wondered if we hadn't noticed them in all our excitement or had they come through while we searched for the bear.

We left one machine on the ridge above the creek and made a large circle of the area determined to retrieve the meat. We found a patch of freshly trampled grass in one area but no blood. It may have been where a moose had bedded down. We continued to search throughout the morning but found no other signs of the bear. We had lost the bear!

We found fresh flagging tape on tree branches when returning to the machine. They continued from our claim markers. Bob thought the intruders had marked their prospecting claim lines. Now we really questioned. Who filed first? Them or us?

"I don't recall any flagging tape in June when building the trail to the creek," I said. "When were they here?"

We thought the bear may have headed for water so we checked the ridge between two small feeder creeks. Nothing! While Bob filled the water containers he told me, "I had read, when an animal is wounded, it will head to the nearest water."

"Hummmmm!" I don't think we will use the water for drinking. Just to be on the safe side." I remarked

Team Work

We returned in the afternoon to finish the hole I had started yesterday by the quartz rock. It became rather difficult to pry the rocks apart where they were wedged together. I went to search for Bob when my wrists became sore from jamming the shovel to pry the rocks apart. Besides, I wasn't gaining much depth.

He too, had only a two and a half-foot hole dug because of the difficulty in digging. He explained, "I was taking it easy. My back is bothering me a lot."

We worked together. Bob used the pick to loosen the rocks and gravel while I shoveled the loosened material out of the hole. We broke through the layer of rock at three and a half feet where Bob took samples. We continued to a depth of five feet, which may not seem like a large hole, but digging through the rock layer was rather difficult.

I realized how tired my body was while hiking back to camp. I looked at Bob, who also looked tired .

"Do you realize what we have done today?" I asked. We searched for a wounded bear and dug holes. I am tired! It's time to return home."

Another Bear

We picked blueberries shortly after leaving Camp #2 and filled four large coffee cans. We packed the berries in ziplok bags and nestled them on our jackets in the boxes on the four wheelers to cushion them for the long eighteen mile ride over the rough trail. We hoped they would survive.

We thought we had an exciting bear story to tell our company. Josie, Bob's sister, interrupted as we related our story to them. She proceeded to tell about their experience. She apologized for breaking in, but she couldn't wait for us to finish our story. She proceeded to tell about their experience.

"You don't have to go way out for a bear! There was a black bear right on your porch about midnight last night!" she excitedly said. Then added, "I heard a loud crash and got up to investigate but without my glasses on, I could only make out a large black animal on the porch. I thought it was a dog but realized it was much bigger! I woke Bob (her husband) so he could see it. Neither one of us had our glasses on and the camera was in the truck. It probably smelled the kielbasa we cooked that evening. It knocked the grill down and was licking it!"

"Then!" she added, "The bear started to come towards our camper. I had this vision of it clawing at the sides and getting in! It stopped in front of the pickup, looked around and headed down the river trail. Now we wondered! Is Ted all right? He leaves his camper door open. Then, were the two chickens and Fred, the rooster, safe!"

"Well, Fred crowed a little late this morning," she laughed, "Let me show you what the bear did to their yard."

She pointed out the bent door to the chicken yard and the damage around the small opening to their shelter where the bear had tried to reach them.

Then she pointed to the imprint of a bear paw in the soft ground near the chicken yard and said, "Look at this! He dragged the bag of grain from the shed to under the picnic table and had a picnic! But he didn't have to spill so much along the way. You had better get him before he gets Fred and the girls!.

"At least they had four feet inside to keep away from the opening," Bob remarked.

Bob straightened the chicken yard and cleaned up the grain. Then he strung a rope between the porch and a tree to hang a pail containing salmon and a few store bought chicken bones. For an alarm system, he attached a cowbell and cans to the rope in hopes it would alert us should the bear return. Everyone slept lightly that night, especially Bob and Josie. They stayed up most of the night hoping to get a picture of the bear. They had their glasses on and camera ready. No bear came.

Bears Return to Camp #3

We set out for the mining claims, two days after everyone left and felt free! We made good time and four hours and forty-five minutes later, arrived at our "Lucky Acres" mining cabin.

Oh no! Bears again! The shovel was twenty feet from where it had hung on the cabin wall. The hammer was moved from its perch on a log by the door and laid upon the ground. The boards, with protruding nails placed on the ground around the cabin, apparently kept the bear from scratching on the plastic or climbing on the roof. The rags, left on the woodpile had been played with again. They were wadded up into a tight ball and placed in the center of a circle of freshly disturbed earth. It appeared the bear enjoyed playing with the rags.

It rained all the next day but the cabin felt warm and cozy. I cooked on the stove inside and appreciated the comfort. I thought of all the rainy days we had sat under the tarp with the dampness and cool breeze. I sure didn't miss cooking over the open fire in the rain! The light coming through the window was too dim to read by at 9:15 p.m. Darkness sets in earlier by mid-August.

Several times during the night we were awakened by noises just outside the cabin. I insisted we investigate. Nothing! I thought I smelled rotten meat in the early morning hours and could have sworn I was awake at the time. I mentioned it to Bob in the morning.

"I never smelled anything. You must have dreamed it," he responded

I wondered, "Had I dreamt it?' It seemed so real to me.

Not a Dream

One pail held two inches of rainwater from the rains of yesterday. Donning our rain jackets and hip boots to protect us from the wet brush, we set out to gather more samples. Wherever my dowsing rods crossed along the ridge above the quartz rocks we took a sample. I believe in them. Bob at times, showed doubts but, without another choice, he went along with me.

When I first started using the dowsing rods, he wasn't too enthusiastic. One day while prospecting he wasn't finding any gold. He panned the gravel where my rods indicated and found a fairly good size flake. But to this day, he is still a doubter because they don't work for him. Maybe I will find the lost gold mine with the dowsing rods. Just to prove they do work!

Only one of the twenty samples I found tested for minerals. We returned to the area to dig more test holes. Again, we worked together. Bob picked and I shoveled. The southwest side of the ridge had smaller rocks mixed with gravel and schist but when panned, there was no black sand.

"There are fairly fresh black bear tracks in the raked area," Bob said, while I prepared supper.

To check if an animal had been in camp, we kept the area where we cut sod for the roof raked smooth.

"It may have come through while we were out gathering samples," he said.

"Do you suppose it could have been earlier this morning when

I thought I smelled rotten meat?" I asked. "They say a bear smells like cow manure," I added.

"Could have been," he answered. The dunny seat was found tipped over the next morning. A bear paw print was in the pile of gravel next to it. Did it happen yesterday while we were gone? Or during the night?

More Quartz

On the day we planned to return home, we gathered more samples from the valley below the Dome, north of camp. After a while, we found ourselves fairly high on a steep ridge on the other side of the creek. Slowly, we descended to where we heard water running. We worked our way through the thick brush with some difficulty to the creek. Bob panned along the way and was ahead of me when I spotted a large quartz rock hidden beneath the banking of the creek.

I noticed a small portion of a quartz rock protruding through the moss, while resting on the steep climb back to the cabin. I insisted that we take time to dig around it for soil samples. The rock and gravel consisted of quartz and breccia, which are associated with gold. Again we found no black sand! It was time to return home!

We checked Camp #2 and noted the coals in the fire pit were still warm. Someone had camped there. We wondered who!

The Campers

We again picked blueberries as the last ones survived the rough trip. A man and his son on four wheelers were stopped by the trapper's home-built hunting trailer along the trail when we approached them. They admitted using our camp as they felt protected with their tent set up in the three-sided shelter. They felt safe until a small grizzly came into camp that morning.

Ahh! I thought, "I knew that camp was on a bear trail." They also enjoyed cutting new trails and found our trail to the water source. They followed a flagged trail to a helicopter pad but lost it beyond it. Bob explained it was our flagging tape marking the trail and told them where the water source was located. They had checked the steep trail but had not found water.

When we arrived home six hours later, we discovered that the bear had gotten into the chicken yard. He tore sections of plywood from the building and more insulation from the small opening when he tried to get to the chickens. Lin, who had checked on the chickens that morning, said there was no bear sign.

Fall Time

We decided to combine the next trip to the claims with moose season,which started September first. The trail was a mess and muddy from hunting traffic but we had no problems.

A large, black animal emerged from the woods as we approached Tim and Eva's hunting campsite. It really scared me before I realized it was Tim's dog, Sandy. They were hunting and hadn't seen any moose sign in the area. Bob told them of the bull moose we had seen several times in the lake when traveling to our claims. The moose seem to know when it is hunting season and disappear.

It was obvious that no one had been on the trail recently. The Delta area had high winds the week before and several trees were down and blocking the trail. I noticed fresh piles of bear scat close to Camp #2. We found the fire pit and garbage dug up and although the people who had camped there had left a clean camp. The recent odor of food had apparently attracted the bear.

Bob noticed that the supply barrel was missing. Did someone steal it? Did Tim and Eva take it? No! We found it in the bushes about fifty feet away laying on it side. The bear or bears had tried to get into it but the locking cover held. Bob tied it with strong rope to a nearby tree. "What are we going to find at Camp #3, " I wondered?

It had also been visited by bears. We arrived at 5:15 p.m. after a long day of clearing trail to find the rags in a neat pile in the center of another well worn circle. Nothing else had been touched!

Bob noticed large moose tracks in the yard and wondered if it had been a bull or cow. He enthusiastically said, "It could have been a bull as I noticed fresh rubbing not too far back along the trail. Maybe we'll get our moose!"

I started a fire in the stove to warm the cabin but cooked the grouse that Bob shot on the way in, over the open fire. The cabin looked warm and cozy in the waning light of day with the warm glow of light in the windows from the Coleman lantern and smoke rising from the chimney. Ahh! Home away from home.

A Different Animal

I cooked breakfast inside as it was quite cool, but a beautiful, morning. I noticed fresh bear scat on the trail to the water source where over a month ago Bob installed a water trough. It was still intact. I mentioned sighting fresh bear scat on the way.

His response was, "I noticed fresh moose tracks, not bear scat, on our creek trail." I knew he would ignore my concern over the bear scat. But I didn't forget it!

One of the intruders had called in July and told Bob they probably would be at camp in August. Curious, we headed downstream to check their campsite.

No one had been back to the camp except for the bear but it didn't get into the cache again. There had been a few rags left on the ground and they were now in a tight, neatly wadded pile! I decided it had to be the same bear that comes to our cabin for it sure enjoyed playing with rags!

We returned to camp. After a quick lunch we headed to a second drainage area below the dome. We followed a trail for a short way then veered off at an angle. Down we went through the thicket to a creek where we had checked once before. We took samples as we crossed the creek to climb the further ridge.

I caught a glimpse of a small, black animal as we crossed the creek. It disappeared behind a downed log. When it reappeared, Bob saw it too. He thought it to be a fairly decent size marten. It's fur shimmered as it ran along the log. It stopped often to look us over, then would disappear again.

Another smaller marten was peering at us through the low branches of a spruce tree. It bobbed its head up and down between the branches as though it were hidden. It was the cutest thing with its ears and face similar to house cat's, except for a longer nose.

We were in a drainage area not indicated on the topographical map. Bob took readings on the GPS and samples. The climb out of the valley was difficult in the heat of the day. We came upon the animal trail below the dome and followed it back to camp.

Bob panned the samples from the second drainage and found traces of black sand (magnetite). I set some of the samples aside for later testing.

Bob mixed charcoal and water into a paste to camouflage the ends of the cabin logs. The freshly cut ends stood out since he cut the open area for a view. We showered in the cool evening breeze and retired to the warm cabin.

Bob said, as he checked the topographical map, "The hike today was less than a mile according to the map."

I thought about that for a second and asked, "Did you take into account the terrain in a square mile section?" And added, "It felt more like two or three miles because of the steepness. The map doesn't indicate the terrain in its measurements of a square mile!"

More Prospectors

The morning of September 4th was beautiful with a few clouds overhead. I packed a lunch for a climb up "the hill" on the other side of the water source creek. On the way down the steep ridge to the creek, I could see "the hill" and thought to myself, "Why did you suggest that we climb this way?" My suggestion was to climb the steep side first while we were fresh in the morning and then, after a long day of hiking, it would be easier to come down to the machine.

It was a difficult climb as usual on the soft moss. I carried the pack. Bob carried the .44 revolver and the shovel to dig samples with along the way. Traveling across the top of the ridge through a burn area was rough. We climbed over downed and criss crossed burnt trees. We stopped for lunch in a grove of poplar trees.

We followed the south side of the ridge towards our main trail by the helicopter pad. Bob noticed a bright pink ribbon on a bush. Descending towards it, we crossed the burn area again having to climb over downed trees.

The ribbon was a marker. A large company, MGM, had either taken soil samples or did magnetometer testing. We wondered if it had been a Japanese company for there was a foil packet with Japanese writing and a picture of a fish. Smelled like fish too! Also lying on the ground was a bag of unopened M&Ms, two bags of crushed cookies and a partial bag of M&Ms in a ziplok bag. No animal had touched them. They littered!

We thought a moose trail would be the easiest path to follow but with their long legs they maneuvered over the downed trees much easier than we could. Bob saw the rock slide he had been looking for on the other side of Ahtel Creek. We had seen this slide in 1995 during our first year of searching for the lost gold mine.

" We'll return to the machine, drive it to camp to get yours, then head over to the ridge above the rocks. It will be the shortest way down to the rock slide," Bob suggested as we continued across the ridge before heading back.

I noted, as I followed Bob, that there were freshly packed paths around the abundant blueberry and cranberry plants and fresh bear scat. I thought to myself, "Oh no, again!"

When we returned to the machine the ten grouse we stirred up that morning were still in the same area.We looked at each other, although it was tempting, I shook my head and said, "Nah! We had grouse last night."

We traveled a mile and a half back down the trail between Camp #3 and #2 before veering off into the brush. To locate the machines in the thick brush Bob tied a piece of flagging tape high in a dead tree. I observed where the machines were parked before we left the area. Each time we start out hiking I look for a ridge, or a particular group of trees that stand out or whatever catches my attention. I need to know my way back to the machines in case something happened to Bob.

Following the ridge was difficult through the thick brush. We circled around to the other side of the valley to the rockslide. We noted the rocks appeared different than other ridges as they seemed to be a mixture of rust colored quartz combined with granite. The rocks of the slide appeared to have happened many years ago for they were covered with moss.

While I dug through one section of the slide I pulled a rock out, examined it, then kept the best sample, Bob was off exploring other areas surrounding the slide

An Indication of Man

"I think I found a dam!" he called out. "In the book it mentioned something about a dam to hold water."

It appeared to be a manmade ditch. Or did we read something into it? A two foot mound of moss covered rocks appeared to have been set in place. Below, there weren't any. It appeared the mound created a ditch to catch rain water or spring runoff when the ground thawed. We heard water running beneath the rocks but apparently it was seeping into the ground.

A sharp indent into the side of the hill nearby gave the appearance it could have been a tunnel. It would be difficult to dig out because of overgrown trees and roots covering the rocks.

"It could all be our imagination. Nothing here but a natural rock area. Let's go see what's further above here," Bob said.

We put our soil and rock samples in the small knapsack.

"Wow! Can we leave some, its heavy!" I said as I placed the pack on my back.

"If you want to lighten your pack, take your samples out," was Bob's comment.

"No way! I think they are good samples," I snapped back. All he answered was, "Well."

The pack was heavier than I thought. I wasn't about to complain as we began to climb.

Nothing different on that ridge either. They all looked a like, with brush, moss and downed trees. We had no problem finding the machines, thanks to the tape in the tree. We cut, split and stacked wood throughout the evening. We realized when we decided to quit how tired we were.

Start A New Trail

I packed a lunch the morning of September 5th before we headed out to explore another area across Koda Creek, where two feeders from Pedro Dome joined. Bob maneuvered the machine with both of us on it through the woods behind one helicopter pad.

We traveled about a half-mile on the machine then cut another half mile of trail to a small protruding knoll where we had lunch. It was quite a view overlooking the vast wilderness beyond us. I heard a snort that sounded like a horse. Bob took the shotgun out of its case. We stood quietly and listened. I could hear my heart pounding! Ten minutes later of not hearing anything Bob turned to me and said,

"You're still paranoid about bears! I don't think you heard anything!" he snapped. I know I did!

We found several trees where a bull moose had rubbed its antlers as we worked our way down the ridge. There were places a little further down where moose had recently bedded down. Several animal trails led to the area. We quit in the late afternoon, tired and hot, and planned to hike across Koda Creek on the next trip. Bob felt we needed to explore more areas as we were becoming discouraged.

We picked blueberries and cranberries on the way back to camp. We both had our responsibilities to close up camp, possibly for the winter. Bob prepared the cabin for winter while I made a list of everything stored in the barrels and in the cabin.

It was sad leaving our cabin, not knowing if we would return or not.

Hunting and Sharing

We were successful in hunting and shot one caribou. Then a week into moose season a new neighbor, Tim, shot his first moose out on the mining road across the highway from Tenderfoot Hill. It took five hours for him to skin and load it into his truck. He was all excited as his hunting on Kodiak Island where he comes from consisted of small black tail deer.

It was such a massive animal to him that he shared it with us and a few neighbors. We taught them how to cut and preserve meat without electricity, running water or refrigeration. Hamburg was ground with a hand grinder. We made moose sausage and liver sausage. Brenda and I canned eighty-four quarts of stew meat. Canned soup made from the scrap meat and bones with vegetables from the garden. They smoked and dried their share of the extra meat. We froze our share in our freezer stored at the neighbors with electricity. For them, it was a learning experience that took a week to complete. But we had meat!

I Become Obsessed!

We decided on September sixteenth to take one more trip to our cabin. Bob didn't want me to take Lin's phone but I insisted. I did a lot of thinking on the way in and kept remembering Bob's words over and over again, "You're too paranoid about bears!"

"If I shot one, maybe I would get over the fear if I knew I could handle myself," I thought. Then I kept these thoughts going over and over in my mind. Now I felt I really wanted to see one. Shoot one!

We were at Lucky Acres cabin in four and a half-hours with no problems on the trail. A bull moose had rubbed trees along the trail. I yelled to Bob when he had slowed down. "Maybe we'll shoot one!" I thought to myself, "Maybe, I'll shoot a bear!"

He looked at me over his shoulder and yelled back, "Are you crazy! How would we get it out without the trailer?"

The Other Side

We followed our new trail below the helicopter pad the next morning. We parked the machine on the ridge and started down the slope to Koda Creek. I thought about the large area we planned to cover and asked Bob to mark the trail with flagging tape. He gave me a disgusted look and hesitated. I explained, if something happened to him I would have to get back to the machines. He flagged it as we descended to the creek. Maybe I worry too much but I always want to be prepared for an emergency. Knowing the cellular phone was in the backpack, I felt secure.

The creek averaged three to five feet wide with bankings two to five feet high covered with thick brush. Bob spread eagle across the creek in a narrow section to give me security as I stepped across. I made it!

The climb was gradual at first as we followed the manmade trail to Ben Leclerc's cabin we found in 1995. I carried the pack. Bob the .44, shovel and machete which he used to cut branches. We veered away from the trail and began the steep climb. We rested every two or three hundred feet of climbing on the deep moss covered ground. Although somewhat overgrown with brush, about half way up the ridge it appeared we were following an old trail.

Eva had mentioned she climbed Pedro Dome on horseback several years ago and thought it was a gradual slope following an old army road to the top. This trail did not appear to have been a road of any sort.

Bob offered to carry the pack. We continued to plod on until 1:15 p.m., where we found a good spot to eat lunch. Ten minutes later we were on the trail. No time to waste. Something caught my eye as we climbed higher. I thought it was an animal and kept glancing up at it. Of course I was afraid to mention it to Bob as I knew what he would say.

When I did tell him I could see something different high up on the treeless part of the ridge he used the binoculars to check. While I waited for him to inform me of what he saw, my mind kept whirling, "It's a bear!" I could feel the adrenaline surging through my body.

"I can't make out what kind of rocks they are but they appear to be white." Then quietly asked himself, "Was it quartz? schist?" he turned to me and said, "There are three separate areas we should check. Not today for those dark clouds are moving in awfully fast and it's a long way back to the machine."

"Rocks!" I thought to myself. "It should have been a bear for me to shoot. Ha. I am glad it's rocks."

We began our descent to the right of the ridge. Down a steep slope to a dry creek with large quartz rocks. We followed it down and Bob panned gravel in pools of water finding traces of black sand. There were large granite rocks along each side of the creek as we maneuvered our way down through the thick brush.

Ben's Cabin

Bob stopped suddenly as we approached the lower section of the creek. He looked at me with a surprised look on his face and said, "Look! It's the cabin"

I saw what appeared to be a cabin through the brush and asked, "But I thought we were on the creek where we put a marker for my prospecting claim in 1995? Somewhere, the two feeders joined above the cabin and not at the creek."

I checked the cabin while Bob took a reading on the GPS. He had taken a reading wherever he took samples or panned. I unearthed a stainless steel copper bottom frying pan in the rubble around the cabin. Bob didn't want me to take it but I figured no one had used it in years and I needed one at camp.

We followed the trail back to where we crossed the creek that morning. We decided not to climb the steep hill back to the machine after crossing a swampy area with lots of moose trails. We opted to follow another creek part way up. It too became steep so we followed the ridge back to the machine.

It took us forty minutes to climb from the creek to the machine as it was not an easy trek. I chuckled when I remembered I had asked Bob to mark the trail, which we never followed back. Would I have if I was by myself? I didn't know.

On the way back Bob stopped to check an outcropping of rocks below the ridge we crossed.. I picked blueberries while waiting. On his return he informed me it consisted mostly of schist.

Dusk was now at 8:30 p.m. The temperature was 48 degrees. Bob decided with a dirt floor in the cabin he could set the shower up inside. Plans for the next morning depended on how our bodies felt. We were lame. It was a day of rest WORKING around camp.

Pedro Dome Again

The morning of September 19th was a cold 32 degrees. It would be a good day to climb although we were lame. We again parked the machine on the ridge below the helicopter pad and headed down to the creek. Bob tried to find a way to drive the machine down the ridge but it wouldn't be easy.

We shared the pack. I carried it down hill while he sampled along the way. He carried it up steep sections. The trail we followed was the same as the day before. We continued to travel through the

Figure 8: Pedro Dome again.

thick brush of alders for some distance above the tree line. Then it was a combination of moss and Hudson Bay tea plants which are two to three feet high and miserable to walk through. We discovered, once we emerged from the thicket, that the area of rocks were granite with a few small quartz mixed in.

I used the dowsing rods as we climbed to another outcrop of rocks. I became disgusted when Bob paid no attention to me each time I called him when the rods crossed. Whether he hadn't heard me or wasn't interested enough to check, I became discouraged and I quit dowsing!

High fluffy white clouds moved overhead and blocked the warm sunshine while we ate our lunch. I noticed a pile of old bear scat nearby and mentioned I had seen several piles along the trail.

"Yeah, yeah!" Bob responded, as though he hadn't heard me. While I repacked the pack he climbed a nearby lone spruce tree. He yelled, "There's another outcropping I want to check!"

"Where?" I asked.

He pointed and said, "Over there, across the valley about a half mile."

"It's just more granite!" I quickly responded with a snap. I followed. It was difficult following him along the edge of the steep ridge with hidden moss covered rocks. I wrenched the muscle in my upper right thigh when I slipped on a rock. It hurt! I sat for a few minutes but knew I had no choice but to continue.

I felt secure with the cellular phone in the pack until Bob pointed out where we were.

"Even if a helicopter came to rescue you, how would they get you up the steep ridge?" he smugly said, implying that the phone was useless.

I looked up the ridge, looked at him and sheepishly said, "They could use a sling from the helicopter like you see on television."

We continued on a few minutes later but I lagged behind. When I approached the rock outcrop I angrily snapped, "Only granite rocks! Just like I said they would be! Now, we have to cut back across the half mile valley again!"

The pain in my thigh eased somewhat as I followed him through the three foot high alder brush, moss and rocks.

"How's your leg?" he asked when we reached the other side of the valley. Then he added, "Do you want to go down the ridge we came up or follow the creek on the other side?"

"My leg feels better although it is still tight. We are this far, so let's go down the creek," I answered.

He looked pleased with my decision and offered to carry the pack. We traveled a short distance down a forty-five-degree incline to the thick brush along the creek. Bob stopped several times to pan but found only minute amounts of magnetite. It was miserable as we worked our way through the thick brush. I became aware of a well used path through the brush but didn't say anything. I thought Bob noticed it too, but he didn't say anything.

Ice coated some wet areas where we crossed Koda Creek. Dark clouds began to form overhead. The last one mile climb to the machine took fifty five minutes. What a blessing it was to ride the rest of the way back to camp.

Although the sky looked threatening we stopped to pick blueberries until it began to hail! We were quite damp by the time we arrived at the cabin. It turned very cold that evening. We decided it was time to close up camp for the winter.

The temperature hovered at thirty-seven degrees when we started for home at ten o'clock on the morning of September twentieth. It was a cold four and a half-hour ride back to the vehicle.

We felt sad having to leave our little cabin until next year. I was satisfied that at our age we could still climb steep ridges and twice climbed Pedro Dome. But very disappointed we hadn't found the lost gold mine. A saying we have read goes like this:

You never know what's possible until you challenge it!

CHAPTER FIVE

Our Fourth Year

Spring of 1998 was earlier than usual which made it possible for us to complete a few main projects and to prepare for company.

Bob asked after a long workday if I could be ready to leave in the morning for an overnight trip to check on the cabin. I was tired but thought a change of scenery would do us both good. Besides, I also, was curious if the bears had gotten into the cabin. The first thing I did was call Lin to borrow her cellular phone. Bob didn't want to be bothered with it just for an overnight trip. But I was still paranoid about bear encounters.

We expected a lot of debris on the trail after the long winter and were prepared for a long trip. We left Fish Lake at eleven thirty a.m. on June third and made good time over the trail around the lake because the mud holes were dry.

The trail had been cleared by someone as far as Squaw Peak. Probably to view the thirty seven thousand acre forest fire burning in the Delta River valley at that time. It was caused by a lightning strike on May twenty first and over twelve hundred people worked on the fire. Bob cleared more brush to get a better view of the fire which was still burning and covered many miles across the vast valley.

It was an easier trip than we had expected with very little debris to clear from the trail. Squirrels along the trail would noisily chatter at us as we passed by. No birds, moose or BEAR SIGN! Camps one and two were fine. We arrived at Lucky Acres cabin sore and tired after six and a half hours on the trail. It too was fine and secure.

The Intruder's Camp

We managed to fill two containers with water at the water source. It was barely a trickle. Surprisingly, the trough Bob had set in the creek was still in place. I do bring drinking water from home on each trip. We hiked the short distance from the water source to the intruder's campsite, located a short distance downstream. A lonely rag, rolled into a tight wad lay by the fire pit. It had been partially buried by the bear.Most of their supplies had been removed. Had they given up their search for the lost gold mine?

I used the stove inside the cabin to cook on. The woods were dangerously dry because of a drought in the area.

Short trip. It was a miserable wet morning but we had to return home. Traveling in the rain, wearing rain gear was not enjoyable. In fact, total misery! We were back at the lake in less than four hours. We had no plans to return until our expected company left.

Free Again

Company had come and gone and within a few days of recuperating, cleaning yard and house we were again on the trail. Although we had received an inch of rain the trail had remained dry. Rion Lake, with not a ripple on its surface appeared as smooth as glass. A small family of ducks swam near the beaver house close to the shore we stood on. A lone Loon, with its reflection shimmering upon the water, was at the far side of the lake.

The trail was ablaze with the many colors of the blooming wildflowers. Patches of bright fuchsia pink fireweed dominated over the numerous other shades of the purples of the monk flower and irises, the white of the flowering dogwood and yarrow, with an occasional daisy gracing the trails edge. The colors contrasted beautifully against the lush green of the grasses and leaves. Oh, oh, a huge pile of bear scat next to the trail near Squaw Peak. Thank goodness it appeared to be several days old.

The prolonged drought of Interior Alaska had an affect on the berry crop. A few berries through the burn area had begun to ripen but the majority of the crop looked poor. We had only one and three quarters of an inch of moisture since January with the normal being three inches by early July.

The cabin was intact upon our arrival four and a half hours later. The rags were on the ground shredded and scattered about not in a tight wad as before. I thought it had been done by the bear but Bob thought a porcupine could have done it.

Bob gathered a few wild Boletis mushrooms growing near the fire pit to add to our supper of moose steaks cooked over the open fire, mashed potatoes, and combined fried zucchini and mushrooms. Yummm!

I left the dirty dishes to conserve the drinking water until we checked the water source. When unpacking supplies from the barrels I noticed something had dug around the half barrel used for panning. Was it a bear? A porcupine? A squirrel? What?

I was exhausted and had fallen right to sleep. Suddenly I was startled awake by Bob giggling. I heard the gnawing sound on the outside wall by my head.

"What is it?" I quietly asked.

"Were you asleep?" he asked, "It's a squirrel," he said as he got out of bed.

It would hesitate each time I banged on the wall but moments later would continue to gnaw. Bob went out to scare it away.

"Too bad we don't have the twenty two with us. It could become a pest," I mentioned, as I followed him out.

Mother Nature called at two thirty in the morning. In the barely diminished light, of night I noticed the sky to the east glowed a bright crimson red. To the east northeast, what appeared to be a full moon was partially obscured by a large spruce tree. A bright white glow barely visible over the Tanana Hills indicated the beginning of the early morning sunrise within an hour.

We Dig Again

I began to pack our lunch after the morning chores for a day of prospecting. I removed the pail from the hole we had dug in one corner of the cabin. I saw something move and thought it was dirt from the sides of the hole falling into the bucket. A little creature looked up at me when I lifted the bucket out of the hole. It dove down between the food to the bottom. I carried the bucket outside and it struggled to climb over the rim. It jumped to the ground and reentered the cabin through a small opening. It was a fairly large shrew which is a small mouse like animal but larger than a gray house mouse. I kept the cover on the bucket after that episode.

We traveled from camp on one four wheeler for about one and a third miles. We parked the machine a short way off the trail then hiked another three quarters of a mile to explore the rock pile we had found last fall on our last day on the claims.

We rechecked below the rock pile what appeared to be a

manmade ditch. Water ran beneath the rocks but we could not pinpoint as to where it flowed.

I cleared moss from what seemed to be the wall. The larger rocks did not appear to have been set in place. The small fine rocks uncovered were of all sizes, shapes and types. I thought it could have been their crushed tailings. Bob had no comment.

In the meantime, he began to dig a hole at the base of an indented area against the hillside. Was it a mine entrance? When his back gave out I continued to dig. I soon realized Bob had disappeared. Although the revolver lay on a rock several feet away I was very conscious of being alone. I did become oblivious of my surroundings while concentrating on picking and shoveling until I heard a rustle in the bushes just above me.

My heart stopped!.

I held my breath and listened.

"I hope it's Bob!" I thought, "Should I move to get the gun or stay quiet? If it's an animal it may pass by without seeing me."

The sound of the movement in the brush came closer. I held my breath. I was relieved when just above me Bob emerged from the bushes. I didn't tell him how scared I had been as he would have told me I was paranoid.

He had a handful of dry animal droppings and asked, "Do you know what kind they are?"

"Moose!" I confidently answered.

"Nope, guess again," he said.

Now I questioned, "Caribou?"

"Nope! porcupine," he smugly answered, "I found an area where it lives, and it craps just outside it's hole."

He must have felt guilty for leaving. He reached for the shovel. He was disgusted when his back went out again from shoveling and picking in an awkward position in the narrow hole. I told him I didn't mind doing the work as long as I was capable. We gathered samples at different depths to test later when it became impossible to swing the palauski.

We had to admit there was no indication of any human behavior through that immediate area. Apparently we read something into it in hopes of finding the lost gold mine.

Bob showed me the porcupine hole and droppings when we climbed to the ridge above where we worked. It was a two mile hike back to the machine.

Another Bear

I began to write in my diary while waiting for the campfire to burn down to coals. A movement behind a tree where the trail enters camp caught my eye when I turned to say something to Bob. I realized it was a black bear coming towards us! I jumped up from the chair and tapped Bob on the shoulder. He was sitting sideways to the bear. I couldn't speak! I muttered, "Ah, ah," then pointed to the bear! He turned his head, saw the bear, and quickly jumped up to get the rifle out of the gun case on the four wheeler behind him. I ran into the cabin for the ————SHOTGUN!

The bear looked to be about three hundred pounds. It hesitated a moment while we scurried around for the guns. I rushed out of the cabin with the shotgun just as Bob shot the bear that was within thirty feet of us. Thud! we heard the bullet hit. The bear made a loud growling sound then rolled over and over in the same spot turning its head as if it were trying to bite something.

I excitedly shouted, "Shoot! Shoot again!" while I sighted in on the bear with the twelve gauge pump shotgun. I nervously pulled the trigger. Nothing happened! In my excitement I had forgotten to release the safety.

Bob seemed to hesitate on a second shot. I angrily yelled, "I thought you said the gun was ready to shoot!" Then I realized the safety was on. We had loaded it the night before.

When I shouted, the bear seemed to get his feet under him and within a matter of seconds ran twenty five feet into the yard. It disappeared into the brush directly across from us. It stopped. Hesitated. Then slowly moved another twenty feet or so and sat down. It turned it's head to look back at us.

Two double sonic booms echoed loudly throughout the valley as Bob cautiously moved into the brush. I think it scared the bear as much as it did us. The bear disappeared and so did Bob!

I nervously waited at camp with my back against the four wheeler next to the cabin.

Gun in hand!.

Heart pounding!

I anxiously waited to hear a shot.

Bob returned a few minutes later. He couldn't find any tracks or blood trail to follow. We retraced the trail from where the bear had entered the brush. There was a long six foot strip of a reddish substance containing green berries.

Bob thought it was from a gut shot wound but admitted I was right when he said, "We scared the crap out of him!"

We found the trail of blood and tracks where the leaves were disturbed by the retreating bear. We continued to follow it until it too, like the last bear, totally disappeared. Wow! Another lost bear and in the very same area.

That evening while we sat by the campfire Bob said, "Maybe it's the old prospector or prospectors trying to tell us where to search for the gold. But we keep wounding them."

We Climb Again

The morning of July fifteenth was a warm beautiful Alaskan morning with sixty five degrees by nine thirty a.m. We set out, each on a four wheeler towards Camp #2 on the trail someone had cut towards Pedro Dome last fall. The one point one mile trail had been extended another three tenths of a mile where it ended at a clump of brush. Bob flagged the area where the machines were parked and we headed for the mountain.

We came upon an animal trail we recognized as the one we had followed on one of the supply trips with Tim and Eva in 1995. The trail had been rough cut and someone had driven over it on a four wheeler. It ended abruptly at the base of the lower ridge of the mountain.

The short climb up the steep ridge in the heat was tiring. We enjoyed the view while we ate lunch. The Delta River off in the distance was barely visible through the smoky haze caused by forest fires in the Yukon Territory of Canada and Tok Alaska.

A bulldozed cut through a pass was visible across the valley below us. It was a remnant of the ice road built last winter which followed the winter trail route to a new mine up the Tonsina River.

We crossed the rock strewn plateau dotted with many alpine wildflowers to descend the head waters of Koda Creek. The thick alders along the creek were difficult to walk through.

After taking a few samples where the creek began Bob wanted to follow it further down. I convinced him it would be difficult through the thick brush and we still had to climb back up the steep ridge to the machines. Luckily we came across a rough animal trail to follow up the ridge. We rested and drank often. It was hot!

We came upon the rough cut trail a short distance into the brush atop the ridge we climbed that morning. We followed it to the area we left the machines and couldn't find them! I soon spotted them after zigzagging several times. They were barely visible behind a clump of scrub birch bushes.

While resting back at camp we discussed what our thoughts and questions were.

Where could the mine be?

Is it only a story?

What was the geology of the area?

What had occurred in the area in the last ninety eight years?

Where was the tree line then?

How big were the trees?

The trees now are not large except for a few spruce trees along the creeks. There was no evidence of burnt stumps or trees except from the forest fire of several years ago. Visible evidence of a mine shaft or a camping area would have been covered by moss.

That evening I read in an old geology book we had at camp that gold is not associated with milky white quartz. We wondered why we had seen gold in milky white quartz rocks at museums. It was an old book and may not be accurate compared to the new technology of today, I hoped.

Our buddy the squirrel returned again at six a.m. to gnaw on the outside wall by my head. He ignored the sound of my banging on the wall and continued his usual fifteen minutes of gnawing before he quit for the day.

A New Sign

We decided to take soil samples below the quartz rocks where the ridge dropped down rather steep, we had not yet tested that particular area. It was slow and enjoyable for we were lame from the climb of yesterday.

As we worked our way across the ridge we found a depression in the ground. The moss covered rocks at the outer edge of the depression appeared to have been moved. By man? Or animal? We hoped we found something man disturbed!

We dug for three hours in the afternoon with Bob loosening the rocks and gravel with the pick while I shoveled the dirt out of the hole. There was a small layer of charcoal at the depth of three feet which appeared to be the remains of a campfire. We dug to a depth of five feet where it still appeared to have been disturbed by man. Although we found no evidence of bones, might it have been the grave of the prospector who had died? We gathered soil samples at different depths and several interesting rocks.

I wanted to continue digging but Bob, with his back bothering him, wouldn't let me.

"I feel guilty." he admitted.

"But I don't mind. Its not your fault. Its not like having a lazy man letting me do the work," I tried to reassure him.

It began to sprinkle by the time we returned to the cabin. Thunder rumbled over the Tanana Hills. I cooked over the open fire but we ate inside the cabin where it was warm and dry. The rain continued and the temperature dropped. We showered inside taking advantage of a dirt floor.

A steady rain continued on the morning of July seventeenth. An inch of rain was in the rain gauge Bob had mounted on a stick.

We closed up camp and headed for home wearing rain gear and hip boots. On one of the few breaks I asked Bob, "Did you notice the five piles of bear scat the first five miles from the cabin?"

"Yes, there are more around," he answered.

We made a record trip back in just under four hours! What will we check on our next trip?

More Communications

The weather cleared by mid morning on August second so we headed to the claims. There were several piles of bear scat near

Squaw Peak but nothing more until we were within two miles of the cabin. Oh no, I thought, I hoped we had gotten rid of them.

Everything was intact when we arrived at six fifty p.m. Bob immediately set up a radio system. He used a car radio and a CB radio he found at the recycling center. He strung an antennae wire to a tree about sixty feet from the cabin. Then ran a wire from the battery system of the four wheeler through the cabin wall to the radio and CB on the shelf built especially for them. I was surprised we received the Glennallen radio station, KCAM, located one hundred and thirty miles to the south of camp. Although we heard people talk over the CB, they could not hear us.

The next morning, we set out to gather samples on the north-east side of the ridge. It was drizzling and fifty-two degrees. We came across another area of moss covered rocks near a gully that appeared to have been moved by man. When we tested and panned the samples that evening we found nothing.

Bob metal detected while I finished testing the samples. He found the dowsing rods I left behind when the bear chewed the seat on the four wheeler. Two of the samples showed mineralization. Excited, we headed back to the area the samples came from to gather more samples. We continued to a small feeder creek where I took water samples. It was too shallow to get a decent sample. We agreed to continue down to the main feeder creek where we took soil samples last year.

The water was seeping underground so we continued further downstream. When we came upon running water again, it was too shallow to get an accurate reading.

We turned back upstream to find the quartz rock we found hidden amongst the muck and grass in the creek last year. Somehow we missed it. We gradually climbed across the valley and came upon a trail Bob had cut with the machete on another trek..

We continued to climb towards camp at a fifty or sixty degree grade. It was hard on our bodies. Our legs would shake from the hard steep climb on the soft moss. We thought our hearts must be okay. They sure were taking a beating! No pun intended.

The shortest route is not always the fastest or easiest. We traveled three hours to encircle the valley to the creek and over an hour to climb straight up to camp.

I not only had to contend with Paul Harvey news in the morning, now another station broadcast Lum and Abner, an old radio program from the 1940s, at five forty five. We probably would have to return in time for it.

GOLD!

Another cloudy morning August fifth. More rain predicted. The samples I tested indicated no minerals were present. But the sample material Bob panned had a speck of gold!

In the four years of soil testing that was the first time a speck of gold stayed in the pan behind the magnetite. We could see through the magnifying glass the gold was attached to a piece of white quartz. Although the tested samples showed no mineralization the gold came from the sample near the quartz rock. Remember, we mark the sample number on a drawn map and sample bag, plus tie flagging tape in a nearby tree with the sample number and date.

Neither one of us wanted to dig or hike that day but by mid afternoon we were restless. Bob suggested, "Let's make a trail across the twin domes. That shouldn't be too hard."

"I can take that kind of work. Besides we've wanted to make a trail to the second dome," I responded enthusiastically.

Bob had cut part of the trail through the burn area. We continued to follow the deep rut of an animal trail but straightened the trail in some places. What a blessing to return over the newly cut trail on the four wheeler. Especially in the rain.

Although we weren't expected home until Friday we decided because of the fog and light rain it was too wet to prospect. We headed home. Nothing would be accomplished in the rain.

The ride back was much cooler than last time. The hand and thumb warmers worked and kept our hands warm. We had run the fans on the machines to keep the engines cool on our way in Sunday because of the heat.

Although the trail was still fairly dry I did get stuck in one mud hole because I didn't keep my speed up.

Another Bear

It was raining the morning of August eleventh. We were on

the trail by four twenty and arrived at camp around nine. No major problems on the trail although we each got stuck once. Apparently, a different bear had visited camp. The rag was spread around and not wadded into a ball like it usually was. The boards placed on the ground around the cabin with nails protruding from them seemed to keep the bears away. But the fire pit had been dug out.

I hated to get out of bed when Mother Nature called at two thirty a.m. It was totally dark at that hour of the morning. I woke Bob to get the flashlight which was on the shelf next to his side of the bed. I slept against the wall! It was cloudy, no stars or moon and soooo quiet. Dark and scary. Bob followed me out. He too, mentioned how dark it was. I shined the light of the flashlight around the yard to check for any eyes that glowed. All clear.

I heard around six o'clock the pitter patter of rain. Rain again! Bob started the fire in the stove and crawled back into bed. I guess we don't crawl anymore we now climb back into bed.

We heard a strange sound later that morning while enjoying a cup of coffee in the warm cabin. It sounded like something dripping but the rain had stopped. It had a hollow sound to it. Then it stopped! We didn't hear it anymore and were puzzled as to what it was.

I was curious. With coffee cup in hand I cautiously opened the door to check the yard. Nothing unusual. I walked over to check the thermometer on the tree by the cabin and loudly said, "It's forty eight degrees."

I turned to check the raked area behind the cabin where we cut the sod for the roof. We kept it raked to check if an animal had walked through the area. Just as I took a few steps, a medium size cinnamon black bear stood upright in the bushes about one hundred feet behind the cabin! It looked at me. I looked at it before I came to my senses and quickly called to Bob,

"A brown bear!" It disappeared and so did I! Bob emerged from the cabin door with the .338 rifle as I rushed to get a gun. He called out, "Get the camera!"

I did as he said. We waited, watched and listened but the bear had disappeared. All of a sudden I had to go to the dunny. Nerves? I asked Bob to stand guard while I used it.

While I packed a lunch and the sampling supplies Bob said,

"I bet that noise we thought was a dripping sound was the bear chewing on the chicken bones we left by the fire pit last night. It may have been chewing on them as he passed by the cabin. Remember the hollow sound it made?"

"You mean, he probably was right next to the cabin while we had our coffee?" I asked.

We checked the raked area and there were fresh bear tracks through it! "Most likely. It looks that way," he answered.

We left around eleven o'clock when the clouds began to break away. Both on one four wheeler we drove across the burn area towards the domes and veered off to the northwest to the end of a short trail where we parked the machine.

The countryside had begun to look familiar as we hiked towards the quartz rock again. Each time we flag and trim a trail we end up exploring another area. So much for marked trails. There's a joke about the only legal litter in Alaska is flagging tape. So many mark their trails.

We were thrilled with the discovery of a speck of gold. We crisscrossed the valley in the vicinity of the quartz rock and took twenty three samples. We returned to camp five hours later and realized how hungry we were. Bob picked a different mushroom then we normally pick although it looked like one we had eaten before. But one never knows and we will see what happens.

I tested fifteen samples during the early evening hours until it became too dark inside the cabin. We checked the rocks we had collected that day using a 100x magnifying glass. It looked like specks of gold on some rocks. But?

It was cloudy and drizzly at eight fifteen and forty four degrees. The single mantel Coleman lantern, one of those recycling center specials, was not working properly. I read by candlelight while Bob listened to Charley McCarthy, another old time program on the Nenana radio station.

A Pattern Forms

The next morning I finished the samples. A pattern formed in the location of the samples which indicated minerals were present. After lunch we went back down to the feeder creek to search for that quartz rock again. Bob doubted I would find it but I did as soon as we reached the creek.

He dug out the muck and grass which surrounded the rock. "I wish I could help you," I said.

"You can! Stay out of my way!" he snapped as he continued to work.

I went above where he worked and used the dowsing rods. The drainage area was the only place they indicated anything but not necessarily in the creek itself.

When I returned he had moved down stream about twenty feet from where he had begun to dig. So much for his belief in my dowsing or what I say or think.

I started to pick at the gravel below the quartz rock in the creek. He said he was going downstream to the mouth of the small feeder creek we had checked before. When he returned I panned the gravel and found a minute amount of magnetite. Nothing to get excited about.

We climbed another steep fifty or sixty percent grade as Bob wanted to check another area before we returned to the machine. It was around about way up and over more moss!

We picked blueberries and cranberries before returning to camp. Upon our arrival Bob said, "Oh oh, we had a visitor."

"What makes you think that?" I asked as I looked around. He pointed to the gas can tipped over. It had been bitten and the two remaining pints of gas had leaked out. Normally Bob would put the gas can in the barrel but for some reason he didn't that day.

The fire pit had been dug out and had a very prominent bear imprint in the sand plus scratch marks remained in the base of the pit itself. I wondered if the same cinnamon black bear of yesterday had returned.

I cleaned the berries while Bob went after water. He had seen a large bull moose on the trail close to the creek. He told me to leave the dirty dishes in the kitchen area outside which may attract the bear. "Would we hear it," I wondered. I have slept soundly with the security of the four walls around me. Bob got the lantern working and I read until late.

The Sluice Box

A beautiful morning greeted us with forty seven degrees. I

hung the blankets on the clothesline to air. Cleaned the rugs and cleared the area in the cabin. No broom, I used the rake to level the ground inside. Housework is never done.

Bob put a small sluice box together and filled out mining claim papers. We planned to stake more claims where we found the minute speck of gold. One rock interested him and he asked me to check it with the magnifier. It definitely looked like gold between the crevices in the quartz rock.

We were excited and headed back to the creek We drove to the further dome to put a mining claim paper on the northeast corner of the claim. Remember, each claim must have a claim paper with a persons name and address and the location of the claim as to the township and range on a quadrangle of a topographical map of the area. Bob cleared brush around a birch tree. Placed the papers in a small plastic bottle and duct taped it to a tree. He marked the direction of the claim with, what else! Flagging tape.

We carried the sluice box, pails, pick, shovel, machete and a screen, which was our shower base, down to the creek near the quartz rock.

While he cleared an area of brush and muck to set the sluice box up, I continued to dig in the creek. Unfortunately, since yesterday, the water flow was slower and did not wash the gravel or silt thoroughly. He managed to run a five gallon bucket of gravel through. I found little magnetite when I panned the concentrates.

He went up to the quartz rock further up stream and brought back gravel for me to wash in the ice cold water. It proved to be mostly fine mica schist with only a slight amount of magnetite.

Disappointed and disgusted with our prospects for the gold Bob asked, "Would you mind driving out tonight as they predict rain tomorrow? Its no fun to drive in the rain."

"I think that's a good idea," I answered.

We returned to camp at four thirty. Closed it up and headed home a half hour later. No special events along the trail except we met two men at the junction of the by pass trail. They were working on the trapper's cabin purchased by the man who had recently sold his claim to a large mining company on the Tonsina River. We arrived home at ten o'clock. What a day it had been. Time for caribou hunting.

Miserable Weather

We attended a seminar at the University of Alaska Fairbanks given by Dr. James Madonna, Professor of Mining Extension at the University on Geo-Chem prospecting. Jim taught the classes we took at U.A.F.

Then on Saturday, September twelfth we were on the trail to the claims by 2:20 p.m. The trail had been well traveled over by moose hunters. At first it felt warm but became quite chilly through the timbered areas. Bob shot four grouse along the way. Then at seven o'clock he used a stick to kill his first porcupine. We decided it was a fur bearing animal. He pulled the quills for my earring crafts before dressing it. It weighed at least twenty pounds. The sky looked threatening and it was getting colder.

Trailer Problems

We added the weight of the meat to the trailer which had several 2x6's, a duffel bag with sheets and blankets and miscellaneous items. The trailer unhitched because of the unbalanced weight, or so we thought. It unhitched again within two hundred feet. Bob examined it closer to discover the hitch was worn. He hammered it to fit tightly but after a short distance it again unhitched. That time he tightened the bolt and had no more problems.

It began to sprinkle just as we arrived at the cabin. Although the bear had played with the damaged gas can everything else was intact. I hurried to get the bedding and things needed for morning out of the barrels. Bob started the stove.

We put the fresh killed meat into a barrel to keep it cool and away from, you know what! Just as we got settled the rain began to fall quite heavily.

Bob replaced the generator on the dump special Coleman lantern. I read by a much brighter light.

A Day of Rest

It was 35 degrees in the morning, foggy and a light rain. I cleaned the meat and cut the porcupine into quarters and cleaned the quills. The gray jays appeared looking for scraps of meat. They almost landed on the table next to me but were cautious for they were not used to people.

Bob returned from his walk and said he had seen fresh moose tracks and rubbings along the trail. He suggested we make a stand as it was moose hunting season.

The pot roasted porcupine was delicious! It was a mixture of beef (dark meat) and pork (texture and juiciness). It had a mild flavor probably due to its feeding on berries in the fall. Bob thought it would have a stronger flavor when chewing on bark in the spring.

Oh No, SNOW

We awoke to a heavy wet snow fallen on the morning of September 14th. 33 degrees. It was only a squall, lightly coating the ground and trees. It was damp and cloudy and a light rain fell throughout the morning.

The fog finally lifted about 12:30. We gathered rocks to have assayed for gold and other minerals. Then headed to the water source to check for moose. Fresh tracks but no bull. We drove back to the area where Bob had seen the rubbings. It was overcast and chilly. Chilled after making a stand for two and a half hours the warm cabin was a welcome sight.

The weather forecast reported on the radio was for snow in the higher elevations by morning. We decided to chance it and stay overnight. We could be snowbound by morning!

The next morning was a winter wonderland! The snow clung to the trees and grasses. It was 30 degrees! We were thankful it was only one inch of accumulation on the ground.

The weather forecast on the 8:00 a.m. news was for fifty percent chance of rain tonight and tomorrow in Fairbanks. We wanted to stay but with the wet bushes we would get soaked walking through them. Bob said, "We have a forty-five degree slope to get up Squaw Peak. We should think about heading out. I'll go gather more rock samples while you start closing up camp." We hated to leave knowing it would be the last trip for the year.

The temperature was 34 degrees and cloudy when we left at 1:05 p.m. In other words it was cold! We hoped we would be hauling moose meat back in the trailer but instead we cleaned up fuel cans and garbage at Camp #1.

The trip out was uneventful but cold. We were prepared for the cold. Bob wore insulated coveralls under his rain gear. I had layers on: long johns, turtle neck jersey, shirt, ragg wool sweater and my usual brown jacket all under my rain gear. We both wore warm boots and wool socks. I was quite comfortable but, the hand warmers and thumb warmer on my machine were not working properly. My fingers were cold.

Bob duct taped plastic covers on the handle bars to protect our hands from the wet brush. It worked! Some areas along the trail had more snow then other sections. We made Squaw Peak okay although snow coated the sides of the trail.

I wondered what Bob wanted when he stopped his machine as we crossed a ridge and walked towards me.

He pointed across the vast valley and said, "See that knoll? That's where we make Camp #4 next spring. It's about four miles down and across the valley to the base of the steep knoll. We will hike most of the way packing supplies for the new camp. It's impossible to cut trail across the valley bottom. That's where we continue our search for the lost gold mine."

I looked at him in disbelief, "Are you crazy!" I said, "We'll be another year older! I'm not sure I will be capable of hiking that distance with a heavy pack on my back. We may not be up to it next year."

"But," he said, "Remember the last portion of the poem, "Alaska Challenge" written by Jim Frey, Sr. the old prospector in the Nebesna area. It went like this:

There's lure in the mountains
    and tundra.
The secret's the oil and gold.
I'll hurry back and find them
    before I grow
    ...too old.

We've got to continue searching for the lost gold mine before we both become too old!" he said in a pleading voice.

In my mind I was thinking of another of Jim Frey's sayings that went something like this:

"He thinks we're four feet from a million dollars and I feel like we're a million feet from four dollars."

Bob quietly said, "The ghosts of the old prospectors won't let us find the mine until the year 2000. That will be the one hundredth anniversary when gold was first discovered in the area we are searching.

"Don't you enjoy a challenge in life?" he asked

I felt we had met a challenge in life the past four years through our adventures and bear encounters but I am determined to follow the love of my life. To experience new adventures in our future years together. Even if I have to hike four miles with a pack on my back to explore a new area.

Our new challenge will be to continue the search for the lost gold mine in 1999.

Before we get too old.

DEDICATION

To Bob,

My best friend and husband, I would like to tell you that I enjoyed every minute of being in the wilderness with you. Bear encounters. Forced prospecting climbs. The digging of many prospect holes. But most of all the loving times.

Thank you.

Your Life long partner
Florine